Lecture Notes in Mathematics

Edited by A. Dold and B. Eckmann

729

Ergodic Theory

Proceedings, Oberwolfach, Germany,
June 11–17, 1978

Edited by
M. Denker and K. Jacobs

Springer-Verlag
Berlin Heidelberg New York 1979

Editors

Manfred Denker
Institut für Mathematische Statistik
und Wirtschaftsmathematik
der Georg-August-Universität
Lotzestr. 13
D-3400 Göttingen

Konrad Jacobs
Universität Erlangen-Nürnberg
Mathematisches Institut
Bismarckstr. 1 1/2
D-8520 Erlangen

AMS Subject Classifications (1970): 28A65, 54H20

ISBN 3-540-09517-9 Springer-Verlag Berlin Heidelberg New York
ISBN 0-387-09517-9 Springer-Verlag New York Heidelberg Berlin

Library of Congress Cataloging in Publication Data
Main entry under title: Ergodic theory.
(Lecture notes in mathematics ; 729) Includes bibliographical references and index.
1. Ergodic theory--Congresses. 2. Topological dynamics--Congresses. I. Denker, Manfred, 1944- II. Jacobs, Konrad, 1928- III. Series: Lecture notes in mathematics (Berlin) ; 729.
QA3.L28 no. 729 [QA313] 510'.8s [515'.42] 79-17368

Printing and binding: Beltz Offsetdruck, Hemsbach/Bergstr.
2141/3140-543210

Dedicated to the memory of

R u f u s   B o w e n

(1947-1978)

Introduction

During the week of June 11-17, 1978, a conference on Ergodic Theory was held in the Mathematisches Forschungsinstitut at Oberwolfach, West Germany, in which mathematicians from Austria, Canada, France, Great Britain, India, Israel, Japan, Poland, United States, and West Germany participated. The main topics consisted of recent research, and significant results obtained during 1977 and 1978 were presented. This volume contains those of the results which will not be published elsewhere. The organizers would like to thank the MFO for their support of the conference, the Springer-Verlag for the publication of these articles, and in particular each of the participants for contributing to the success of the conference.

Göttingen/Erlangen,
15th February, 1979

Manfred Denker

Konrad Jacobs

C o n t e n t s :

Participants:

J.Aaronson (Rennes)
R.Adler (Yorktown Heights)
M.A.Akcoglu (Toronto)
St.Alpern (London)
J.Auslander (Maryland)
A.Beck (Madison)
A.Bellow (Evanston)
J.R.Blum (Tucson)
S.G.Dani (Bombay)
M.Denker (Göttingen)
Y.Derrienic (Rennes)
J.Feldman (Berkeley)
H.Haller (Erlangen)
T.Hamachi (Fukuoka)
G.Helmberg (Innsbruck)
M.Herman (Paris)
F.Hofbauer (Wien)
K.Jacobs (Erlangen)
A.B.Katok (Paris)
M.Keane (Rennes)
G.Keller (Münster)
U.Krengel (Göttingen)
W.Krieger (Heidelberg)
F.Ledrappier (Paris)
M.Lin (Beer-Sheva)
D.A.Lind (Seattle)
B.Marcus (Chapel Hill)
M.Misiurewicz (Warschau)
J.Moulin Ollagnier (Paris)
J.Neveu (Paris)
R.Nürnberg (Göttingen)
W.Parry (Coventry)
K.Petersen (Chapel Hill)
D.Pinchon (Paris)
B.Roider (Innsbruck)
K.Schmidt (Coventry)
F.Schweiger (Salzburg)
C.Series (Cambridge)
K.Sigmund (Wien)
M.Smorodinsky (Tel-Aviv)
L.Sucheston (Columbus)
W.Szlenk (Warschau)
M.Thaler (Salzburg)
J.-P.Thouvenot (Paris)
K.M.Wilkinson (Nottingham)

List of talks

Aaronson, J.:	About transformations preserving infinite measures
Adler, R.:	Topological entropy and equivalence of dynamical systems
Akcoglu, M.:	Pointwise ergodic theorems
Alpern, S.:	Generic properties of measure preserving homeomorphisms
Auslander, J.:	Disjointness in ergodic theory and topological dynamics
Bellow, A.:	Another look at a.s. convergence
Blum, R.:	Pointwise ergodic summability methods on LCA groups
Dani, S.G.:	Invariant measures of horospherical flows
Feldman, J.:	Reparametrization of probability-preserving n-flows
Hamachi, T.:	Fundamental homomorphism of normalizer group of ergodic non-singular transformation
Helmberg, G.:	On ε-independence and topological Rohlin-sets
Hofbauer, F.:	Das maximale Maß für die Transformation $T : x \to \beta x + \alpha (\mathrm{mod}\ 1)$
Katok, A.B.:	Lyapunov exponents, entropy and invariant foliation in smooth ergodic theory
Katok, A.B.:	Generalized rotation numbers for Anosov Flows
Keane, M.:	Bernoulli schemes of the same entropy are finitarily isomorphic
Keller, G.:	Piecewise monotonic functions and exactness
Ledrappier, F.:	A variational principle for topological conditional entropy
Lin, M.:	Weak mixing
Lind, D.A.:	Specification for compact group automorphisms
Marcus, B.:	Topological entropy of some skew products
Misiurewicz, M.:	Invariant measures for continuous transformations on [0,1] with zero topological entropy
Moulin Ollagnier, J.:	A new proof of E.Følner's result: Countable amenable groups have an ameaning filter
Neveu, J.:	On the filling schema and a simple proof of the Chacon-Ornstein theorem.
Nürnberg, R.:	Constructions of strictly ergodic systems which are not loosely Bernoulli
Parry, W.:	Generic properties of endomorphisms
Petersen, K.:	Balancing ergodic averages
Schmidt, K.:	Unique ergodicity
Schweiger, F.:	The "jump transformation" and its applications
Series, C.:	Foliations and ergodic equivalence relations
Smorodinsky, M.:	Bernoulli factors that span a transformation
Sucheston, L.:	Operator ergodic theorems for superadditive processes

ON THE CATEGORIES OF ERGODICITY
WHEN THE MEASURE IS INFINITE

Jon AARONSON

Let (X,B,μ) denote the real line equiped with Lebesgue measure, and let G denote the group (under composition) of automorphisms (invertible, measure preserving transformations) of (X,B,μ) endowed with the topology of convergence :

$$T_n \to T \quad \text{if} \quad \mu(T_n^{-1} A \,\Delta\, T^{-1}A) + \mu(T_n A \,\Delta\, TA) \to 0$$

$$\forall A \in \mathcal{F} = \{A \in B : \mu(A) < \infty\}$$

It was shown in [S][1] that the ergodic automorphisms are residual in G.

Here we prove category theorems for some stronger versions of ergodicity.

Let $T \in G$ be ergodic. Then T is conservative, and, by Hopf's ergodic theorem ([E], p.49) :

$$\sum_{k=0}^{n-1} f(T^k x) / \sum_{k=0}^{n-1} g(T^k x) \to \int_X f d\mu / \int_X g d\mu \quad \text{a.e.} \quad \forall f,g \in L^1 \; \int g \neq 0$$

1 The results in [5] actually apply to the semigroup of endomorphisms of (X,B,μ). The proofs of all the results in that paper quoted here can easily be adapted to G (see the chapter on weak topology in [H]).

Note that Hopf's theorem gives no information on the "absolute" asymptotic growth of the sums $\sum_{k=0}^{n-1} f(T^k x)$. ($f \geq 0$). In the stronger versions of ergodicty considered here, such information is (defined to be) available.

Definition 1

(a) Let $T \in G$ and $\{n_k\} \in \mathbb{N}$, $n_k \to \infty$. We will say that $\{n_k\}$ is *good sequence for* T if there are constants $d_k \to \infty$ such that

(1) $$\frac{1}{d_k} \sum_{j=0}^{n_k-1} f(T^j x) \xrightarrow{\mu} \int_X f d\mu \qquad \forall f \in L^1$$

Here "$\phi_n \xrightarrow{\mu} \phi$" means that for some (and hence all) probability measures $P \sim \mu$: $P([|\phi_n - \phi| > \varepsilon]) \to 0 \quad \forall \varepsilon > 0$.

(b) We will say that T is *homogeneous* if $\mathbb{N}$ is good for T .

If $T \in G$ has a good sequence $\{n_k\}$, then T is ergodic and the sequence $\{d_k\}$ satisfying (1) is defined uniquely up to asymptotic equality. We will call this sequence the *return subsequence of* T *along* $\{n_k\}$ (or the *return sequence* of T if $\{n_k\} = \mathbb{N}$).

Definitions 1 (a) and (b) are applicable to automorphisms of a finite measure space where they both coincide with ergodicity. Thus we only consider automorphisms of (X,B,μ), where these definitions do not coincide with ergodicity (see [A]).

We prove :

Theorem 1 : The automorphisms with good sequences are residual in G .

and

Theorem 2 : The homogeneous automorphisms are dense but meagre in G.

We will need the

Conjugacy Lemma ([S][1],[Kr])

Suppose $T \in G$ is aperiodic (i.e. $\mu(\{x : T^n x = x\}) = 0$ $\forall n \geq 1$) then $C(T)$ is dense in G where $C(T) = \{\pi^{-1} T \pi : \pi : X \to X$ is a measureable, invertible, measure multiplying map$\}$.

The existence of homogeneous automorphisms was established by theorem 4.4 of [A]. Any homogeneous automorphism is clearly ergodic and hence aperiodic. Also, if T_0 is homogeneous, then so is every member of $C(T_0)$. Thus, by the conjugacy lemma, *the homogeneous automorphisms form a dense set in* G .

Proof of theorem 1

Let $T_0 \in G$ be homogeneous with return sequence a_n . As mentioned above, $C(T_0)$ is dense in G . Let :

$$\mathcal{Q} = \{T \in G : \exists n_k \to \infty \text{ s.t. } \frac{1}{a_{n_k}} \sum_{j=0}^{n_k - 1} f \circ T^j \to \int_X f d\mu \text{ a.e. } \forall f \in L^1\}$$

Then, $C(T_0) \subseteq \mathcal{Q}$, and every member of $\mathcal{Q}$ has good sequences. We prove theorem 1 by by showing that $\mathcal{Q}$ is a G_δ set in G .

Let $P \sim \mu$ be a probability measure. It is not hard to show that sets of the form

$\{T : P(\sum_{k=0}^{n-1} 1_A \circ T^k \in [\alpha\ \beta] \in [\gamma\ \delta]\}$ are closed in G where $A \in \mathcal{F}$, and $\alpha, \beta, \gamma, \delta \geq 0$.

Thus, if $\{A_\nu\}_{\nu=1}^{\infty}$ is a dense subcollection of $\mathcal{F}$ (i.e. $\forall A \in \mathcal{F} \exists \nu_n$ s.t. $\mu(A_{\nu_n} \Delta A) \to 0$), and $\psi_n(\nu) = \sum_{k=0}^{n-1} 1_{A_\nu} \circ T^k$, the set :

$$\mathcal{A}' = \bigcap_{k=1}^{\infty} \bigcup_{n=k}^{\infty} \bigcap_{\nu=1}^{k} \{T : P([|\frac{1}{a_n} \psi_n(\nu) - \mu(A_\nu)| > \frac{1}{k}]) < \frac{1}{2^k}\}$$

is a G_δ in G .

This proof is therefore completed by establishing that $\mathcal{A} = \mathcal{A}'$

It is not hard to check that :

(2) $\mathcal{A}' = \{T \in G : \exists n_k \text{ s.t. } \frac{1}{a_{n_k}} \psi_{n_k}(\nu) \to \mu(A_\nu) \text{ a.e. } \forall \nu \geq 1\}$

from which the inclusion $\mathcal{A} \subseteq \mathcal{A}'$ follows immediately.

Now suppose $T \in \mathcal{A}'$. From the above representation of $\mathcal{A}'$ we see that :

$$\sum_{n=1}^{\infty} 1_{A_\nu} \circ T^n = \infty \quad \text{a.e.} \quad \forall \nu \geq 1$$

Thus, T is conservative, and, Hopf's theorem (for conservative but possibly non-ergodic transformations) yields.

$$\sum_{j=0}^{n-1} f(T^j x)/\psi_n(\nu)(x) \to h_\nu(f)(x) \quad \text{a.e.} \quad \forall f \in L^1 , \ \nu \geq 1 .$$

where $h_\nu(f) \circ T = h_\nu(f)$ and $\int_{A_\nu} h_\nu(f)\, d\mu = \int_X f d\mu$.

By (2), we now have :

$$\frac{1}{a_{n_k}} \sum_{j=0}^{n_k - 1} f \circ T^j \to h(f) \quad \text{a.e.} \quad \forall f \in L^1 \text{ where}$$

$$\int_{A_\nu} h(f)\, d\mu = \mu(A_\nu) \int_X f d\mu$$

Since $\{A_\nu\}$ is dense in $\mathcal{F}$: $h(f) = \int_X f d\mu \ \ \forall f \in L^1$ and $T \in \mathcal{A}$

□

Proof of theorem 2

As remarked before, the homogeneous automorphisms are dense in G. We show that the non-homogeneous automorphisms are residual in G.

Let $T \in G$ be homogeneous with return sequence $\{a_n\}$. Then, by theorem 4.1 of [A], a_n is regularly varying with index 1 and, in particular, $a_n/\sqrt{n} \to \infty$. Hence : for no $B \in \mathfrak{F}$ $\mu(B) = 1$ and no subsequence $n_k \to \infty$ does :

$$(3) \qquad \frac{1}{\sqrt{n_k}} \sum_{j=o}^{n_k - 1} 1_B \cdot T^j \to 0 \quad \text{a.e}$$

Thus, if we fix an $B \in \mathfrak{F}$, $\mu(B) = 1$ and let $\mathfrak{C}$ $(=\mathfrak{C}_B)$ denote those automorphisms $T \in G$ satisfying the convergence (3) along some subsequence n_k, then $\mathfrak{C}$ consists entirely of non-homogeneous automorphisms.

It is not hard to see that :

$$\mathfrak{C} = \bigcap_{m=1}^{\infty} \bigcup_{n=m}^{\infty} \{T \in G \ : \ P\left(\sum_{k=0}^{n-1} 1_B \circ T^k > \sqrt{n}/m\right) < 1/2^m\}$$

where $P \sim \mu$ and $P(X) < \infty$, so $\mathfrak{C}$ is a G_δ in G.

Now if $T \in \mathfrak{C}$ were ergodic, then, by Hopf's thorem the convergence (3) would be satisfied by T for every $B \in \mathcal{F}$, and hence by every member of $C(T)$ $\forall$ $B \in \mathcal{F}$ along some subsequence n_k.

Thus, for ergodic $T \in \mathcal{C}$: $C(T) \subseteq \mathcal{C}$. So, by the conjugacy Lemma, the existence of an ergodic $T \in \mathcal{C}$ would imply the density (and hence residuality) of $\mathcal{C}$ in G. To complete this proof, here is an ergodic Markov shift in $\mathcal{C}$:

Let $u_n = 1/(n+1)^{3/4}$ $(n \geq 0)$. Then $\underline{u} = \{u_n\}_{n=0}^{\infty}$ is a (bounded) Kaluza sequence (i.e $u_{n+1}/u_n \uparrow$) and hence ([Ka]), a renewal sequence —null-recurrent since $u_n \to 0$ and $\sum u_n = \infty$. Thus ([C]), there is an irreducible, null-recurrent stochastic matrix $Q = \{q_{i,j}\}$ $i,j \in \mathbb{N}$ with the property that $q_{1,1}^{(n)} = u_n$.

Since Q is irreducible and null-recurrent, the two-sided shift of Q : $T_{\underline{u}}$ is an ergodic automorphism of an infinite measure space isomorphic to (X, B, μ) and may therefore be regarded as an (ergodic) element of G.

It remains to show that $T_{\underline{u}} \in \mathcal{C}$:

By construction, $\exists$ $A \in \mathcal{F}$, $\mu(A) > 0$ such that

$$\mu(A \cap T_{\underline{u}}^{-n} A) = \mu(A)\, u_n \quad \forall\, n \geq 1$$

Hence

$$\frac{1}{\sqrt{n}} \sum_{k=0}^{n-1} \mu(A \cap T_{\underline{u}}^{-k} A) \sim 4n^{-1/4} \to 0 \text{ , so } \exists\ n_k \to \infty \text{ such that}$$

$$\frac{1}{\sqrt{n_k}} \sum_{j=0}^{n_k-1} 1_A \circ T_{\underline{u}}^j \to 0 \text{ a.e. on } A.$$ (In this case, we may take $n_k = k$).

The set on which this convergence takes place is $T_{\underline{u}}$-invariant, and, containing $A \neq \emptyset$, must by the ergodicity of $T_{\underline{u}}$ be X. The Hopf ergodic theorem now establishes that $T_{\underline{u}} \in \mathcal{C}$. □

REFERENCES

[A] J. AARONSON : On the pointwise ergodic behaviour of transformations preserving ∞ measures. To appear in Israel Journal of Maths.

[C] K.L. CHUNG : Markov Chains with stationary transition probabilities. Springer 104 Heidelberg (1960).

[E] E. HOPF : Ergodentheorie. Chelsea (1948).

[H] P. HALMOS : Lectures on ergodic theory. Chelsea (1956).

[Ka] T. KALUZA : Über die Koeffizienten reziproker Potenzreihen Math. Z. 28 p. 161-170 (1928).

[Kr] K. KRICKEBERG : Mischende Transformationen auf Mannigfaltigkeiten unendlichen Masses : Z. Wahrsch. verw. Geb. 7 (1966) p.161-181.

[S] U. SACHDEVA : On category of mixing in ∞ measure spaces. Math. Systems theory, 5 (1971), p.319-330.

Jon Aaronson
Laboratoire des Probabilités,
Université de Rennes
Avenue du Général Leclerc
F-35031 Rennes cedex

A Selection of Problems in Topological Dynamics

Roy L. Adler
Mathematical Sciences Department
IBM Thomas J. Watson Research Center
Yorktown Heights, New York

Several results of ergodic theory strongly suggest analogues in topological dynamics. In this talk we shall indicate some unsolved ones which are related to the notion of topological entropy.

We denote a general compact dynamical system by (X,ϕ) where X is a compact metric space and ϕ a homeomorphism of X onto itself. Let $h(X,\phi)$ denote its topological entropy. Two dynamical systems (X,ϕ), (Y,ψ) are called topologically conjugate, $(X,\phi) \approx (Y,\psi)$, if there exists a homeomorphism θ of X onto Y such that $\theta\phi = \psi\theta$. It follows easily from the definitions that $\approx$ is an equivalence relation and topological entropy an invariant.

We shall be concerned mainly with the following special models of dynamic systems: full symbolic shifts (S^Z,σ) where S is a finite set of symbols with the discrete topology, S^Z the space of doubly infinite sequences of elements of S with the product topology, and σ the shift transformation; subshifts (X,σ) where X is a closed σ-invariant subset of S^Z; subshifts of finite type where $X = S^Z - \bigcup_{-\infty}^{\infty} \sigma^j(C_1 \cup C_2 \cup \ldots \cup C_n)$ and $C_1,\ldots,C_n$ are a finite number of finite cylinder sets; topological Markov shifts where $X = (T) \equiv \{x = (\ldots x_{-1},x_0,x_2,\ldots)\colon\ x_i \in S$ $t_{x_n,x_{n+1}} = 1,\ n \in Z\}$ and $T = (t_{ij})$ is a zero-one transition matrix (a subshift of finite type is always topologically conjugate to a topological Markov shift usually over a different symbol set, and conversely); and toral automorphisms where $X = R^n/Z^n$ and ϕ is given by a unimodular matrix.

I. The Conjugacy Problem. Is there an algorithm for determining topological conjugacy and constructing conjugating maps between specified systems? For subshifts of finite type, R. Williams [Wi] has tried but not yet successfully to reduce the problem to one of solving diophantine equations. For toral automorphisms the program has been slightly more successful. Reduction to a diophantine problem was accomplished in [AP] and [Ar]. Two toral automorphisms are topologically conjugate if and only if their corresponding unimodular matrices are conjugate elements in $GL(n,Z)$. Here we are in a special case of the general problem of determining conjugate elements in a finitely presented group which is the second of Dehn's three fundamental decision problems [MKS, p. 24, p. 168], the first being the word problem. For $GL(n,2)$ it has only been settled for $n = 1,2$. For $n = 2$ it is a problem of solving quadratic equations in two variables and connected with Pell's equation.

II. A Weaker Isomorphism. Greater success can be achieved with the program attempted in I by using a slightly weaker notion of equivalence which we call almost topological conjugacy. In terms of this relation we can prove a topological analogue to Ornstein's isomorphism theorem for Markov shifts [O]. The precise measure theoretic version is to be found in [ASS].

Isomorphism Theorem for Topological Markov Shifts [AM]. Two irreducible topological Markov shifts are almost topologically conjugate (with the conjugating maps constructed by an algorithm) if and only if they have the same period and topological entropy.

In order to give a definition of almost topological conjugacy we need two other notions which we shall leave unspecified for the moment. We need a condition of indecomposability restricting the class of dynamical systems and a notion of negligibility for sets where maps fail to be one-to-one. For indecomposable dynamical systems we say (Y,ψ) is an almost conjugate extension of (X,ϕ) or (X,ϕ) an almost conjugate factor of (Y,ψ) if there exists a negligible invariant set $N \subset X$ and a continuous boundedly-finite-to-one map π of Y onto X which satisfies $\phi\pi = \pi\psi$ and maps $Y-\pi^{-1}N$ one-to-one onto $X-N$. Two dynamical systems are said to be almost topologically conjugate, $(X,\phi) \sim (Y,\psi)$, if they have a common almost conjugate extension. It turns out we have more than one version of the relation $\sim$ depending on how we specify indecomposability and negligibility. First version: we restrict dynamical systems to those supported by at least one ergodic invariant probability measure which is positive on open sets and take as negligible sets those which have measure zero with respect to all such measures. Second version: we restrict dynamical systems to topologically transitive nonwandering ones and take as negligible sets those containing points whose future orbit or past orbit hits some open set a finite number of times. Third version: we take the same class of dynamical systems and choose in each the set of nondoubly transitive points as a single universal negligible set. In the last two versions the negligible sets are of first category besides being of measure zero with respect to any ergodically supporting measure if such exist. In each version $\sim$ is an equivalence relation and topological entropy an invariant, although these facts are no longer such easy consequences. The proof with respect to version one was given in [AM]. The details should be carried out for the others. In any case they all lead to the same isomorphism theorem for topological Markov shifts with essentially the same proof. One might ask which version is best. All three equivalence relations are topological in character, but the first uses the concept of measure. The second yields the strongest relation, while the third seems the most appealing. There may well be others, but one must be careful. For example, if we were to call two topologically transitive nonwandering systems equivalent if they are topologically conjugate with respect to the subspace of doubly transitive points, then we get the same isomorphism theorem for subshifts of finite type; but for general subshifts topological entropy will not even be an invariant.

III. Toral Automorphisms. R. Bowen has proved that hyperbolic toral automorphisms are almost conjugate (all versions) factors of aperiodic topological Markov shifts [B-1, B-2]. Thus we have the following corollary to the isomorphism theorem.

Corollary. Two hyperbolic toral automorphisms are almost topologically conjugate if and only if they have the same entropy.

What about nonhyperbolic toral automorphisms? Using the notion of specification D. Lind [L-2] has shown that nonhyperbolic toral automorphisms are not almost conjugate factors of subshifts of finite type; so the isomorphism theorem of [AM] cannot be applied. Perhaps the answer lies in some kind of generalization dealing with skew products of topological Markov shifts and rotations on the circle. This leads to the following question. Given a nonhyperbolic toral automorphism, does there exist a topological Markov shift with the same entropy? Constructing transition matrices T with specified entropies (Perron characteristic values) is a difficult problem and not much is known. However one can find transition matrices which have entropies equal to those of hyperbolic toral automorphisms by using Bowen's result. Thus we are led to an algebraic number theory question. Given a nonhyperbolic toral automorphism, does there exist a hyperbolic one with the same entropy? Next what about a converse to Bowen's results? When is an aperiodic topological Markov shift an almost conjugate extension of a toral automorphism? If this could be proved for an infinite number of $n \times n$ transition matrices of the form

$$T = \begin{bmatrix} 0 & 1 & . & . & . & 0 \\ 0 & 0 & 1 & & & \\ & & & \ddots & & \\ & & & & \ddots & 1 \\ 1 & 1 & 0 & . & . & . \; 0 \end{bmatrix}$$

whose entropy approaches 0 as $n \to \infty$, it would solve the problem of the entropy infimum of toral automorphisms discussed in [L-1]. It goes without saying that we can ask similar questions for natural extensions of toral endomorphisms and other group automorphisms.

IV. Nonalgebraic Entropy. So far we have only dealt with systems whose entropies are logarithms of algebraic numbers. Can we get isomorphism results for systems whose entropies are logarithms of transcendental numbers? Is there a wider class of subshifts beyond sofic ones [CP,We] for which topological entropy is a complete almost topological conjugacy invariant? I believe a promising line would be to investigate symbolic systems defined in terms of renewal blocks such as those associated with the β-transformation. [P].

V. Factors and Extensions. What are the analogues to Ornstein's criteria for determining when a process is isomorphic to a Bernoulli shift? In some sense the work of Bowen [B-1, B-2] on diffeomorphisms fits this description. Are there criteria for determining when one subshift of finite type is an almost conjugate factor of another? What can be said about the class of shifts which have no nontrivial almost conjugate factors? A factor is trivial if it is topologically conjugate to the extension in question. Does the full 2-shift have nontrivial almost conjugate factors?

VI. n-point Extensions. We say (Y,ψ) is an almost n-point extension of (X,ϕ) if there exists a negligible invariant set $N \subset X$ and a continuous boundedly-finite-to-one map π of Y onto X which satisfies $\phi\pi = \pi\psi$ and maps $Y-\pi^{-1}N$ n-to-one onto $X-N$. Two n-point extensions (Y,ψ), (Z,ρ) of (X,ϕ) are called equivalent if they have a common almost conjugate extension (W,τ) such that the factor maps down to (X,ϕ) commute. In [AM] we prove all 1-point extensions are equivalent. However the problem becomes more interesting for $n \geq 3$. What we are after is the analogue to Rudolph's theorem [R] which classifies k-point extensions of Bernoulli shifts by an invariant which is a certain algebraic structure in the symmetric group on k-points. Rudolph's invariant will be one here also, but is it complete? This result could then be used to study the problem of n-point extensions over different systems. An n-point extension (Y,ψ) of (X,ϕ) is said to be equivalent to an n-point extension (W,τ) of (Z,ρ) if (Y,ψ) and (W,τ), have a common almost conjugate extension which is an n-point extension of a common almost conjugate extension of (X,ϕ) and (Z,ρ) and the factor maps in the diagram commute.

VII. Other Group Actions. Are there isomorphism theorems for other group actions such as flows or Z^ν-actions? As a first step for Z^2-actions one should examine the system of Markley and Paul [MP].

VIII. Krieger's Theorem. What is the relation between topological entropy and the number of symbols needed to represent a dynamical system? Is the analogue to Krieger's theorem [K] on generators true? Namely -- given $h(T),\sigma)$, does there exist a subshift of finite type in (S^Z,σ) where the number of symbols in S is the next larger integer to $e^{h((T),\sigma)}$? There may be a way to use the circles of Williams (See [AM]) to get such a result.

REFERENCES

[AM] R. L. Adler and B. Marcus, Topological entropy and equivalence of dynamical systems (preprint), to appear in Memoirs Amer. Math. Soc.

[AP] R. L. Adler and R. Palais, Homeomorphic conjugacy of automorphisms on the torus, Proc. of Amer. Math. Soc. 16 (1965) 1222-1225.

[ASS] R. L. Adler, P. Shields, and M. Smorodinsky, Irreducible Markov Shifts, Ann. of Math. Stat. 43 (1972) 1027-1029.

[Ar] D. Z. Arov, Topological similitude of automorphisms and translations of compact commutative groups, Uspehi Mat. Nauk 18 (1963) 133-138.

[B-1] R. Bowen, Markov partitions and minimal sets for axiom A diffeomorphisms, Amer. J. Math. 92 (1970) 907-918.

[B-2] R. Bowen, Markov partitions for axiom A diffeomorphisms, Amer. J. Math. 92 (1970) 725-747.

[CP] E. M. Coven and M. E. Paul, Sofic systems, Israel J. Math. 20 (1975) 165-177.

[K] W. Krieger, On entropy and generators of measure-preserving transformations, Trans. Amer. Math. Soc. 199 (1970) 453-464.

[L-1] D. A. Lind, Ergodic automorphisms of the infinite torus are Bernoulli, Israel J. Math. 17 (1974) 162-168.

[L-2] D. A. Lind, Ergodic group automorphisms and specification, (this conference).

[MKS] W. Magnus, A. Karass, and D. Solitar, Combinatorial Group Theory, Pure and Appl. Math. 13, Interscience New York (1966).

[MP] N. G. Markley and M. E. Paul, Matrix subshifts for Z^{ν} symbolic dynamics, Univ. of Md. Technical Report Tr 78-33 (1978).

[O] D. S. Ornstein, Ergodic Theory Randomness and Dynamical Systems, Yale Math. Monographs 5, New Haven and London, Yale Univ. Press (1974).

[P] W. Parry, On the β-expansion of real numbers, Acta Math. Acad. Sci. Hungar. 11 (1960) 401-416.

[R] D. Rudolph, Counting the relative finite factors of a Bernoulli shift. Israel Journal of Math. 30 (1978) 255-263.

[Wi] R. F. Williams, Classification of shifts of finite type, Ann. of Math. 98 (1973) 120-153. Errata, Ann. of Math. 99 (1974) 380-381.

[We] B. Weiss, Subshifts of finite type and sofic systems, Monatsh. Math. 77 (1973) 462-474.

Pointwise Ergodic Theorems in L_p Spaces

Mustafa A. Akcoglu

Let $L_p = L_p(X,F,\mu)$ be the usual Banach Spaces associated with a measure space (X,F,μ). Let $T : L_p \to L_p$ be a contraction (i.e. a linear operator with norm not more than 1) and let A_k, $k \geq 1$, denote the polynomial $A_k(z) = \frac{1}{k}\sum_{i=0}^{k-1} z^i$.

By the pointwise ergodic theorem in L_p we mean the existence of

(*) a.e. $\lim_{k\to\infty} A_k(T)f$ for all $f \in L_p$.

This has been investigated by many authors, starting with the Hopf's generalization [8] of the classical Birkhoff theorem [4] to an operator theoretic setting. The situation for a positive contraction ($TL_p^+ \subset L_p^+$) now seems to be well understood: if $1 < p < \infty$ then (*) exists [1], but it does not have to exist if $p = 1$ [6] or if $p = \infty$. The general case of a not necessarily positive contraction is, however, essentially open, although there are some partial results in this direction, the most important ones being the following theorems.

<u>Theorem 1</u> (Dunford-Schwartz [7]): If T is simultaneously a contraction of all L_p spaces, $1 \leq p \leq \infty$, then (*) exists, for all $1 \leq p \leq \infty$.

<u>Theorem 2</u> (Bellow [3]): If $1 < p < \infty$ and $p \neq 2$ and if T is an invertible isometry of L_p then (*) exists.

If $p = 2$ then (*) need not exist, even if T is an invertible isometry of L_2 (i.e. a unitary operator). This follows from a counter-example of Burkholder [5], which depends, in turn, on a counter-example of Menchoff [9]. Since Burkholder's example is originally given in a different context, we would like to describe it in a somewhat simplified way and indicate why it does not seem to clarify the situation for $p \neq 2$. One of the arguments in [5] can be formulated as follows.

<u>Lemma</u>: Let H be a Hilbert space with a complete orthonormal basis φ_n, $n \geq 1$, and let Q_n be the projection on the subspace spanned by $\{\varphi_{n+1}, \varphi_{n+2}, \ldots\}$. Then, for each sequence $\varepsilon_n > 0$, there exists a unitary operator T and a sequence of integers k_n such that $\|A_{k_n}(T) - Q_n\| < \varepsilon_n$ for all $n \geq 1$.

<u>Proof</u>: Let P_n be the projection on φ_n. We define $T = \sum_{n=1}^{\infty} \lambda_n P_n$, where the complex numbers λ_n will be chosen as follows, always satis-

fying the conditions that $|\lambda_n| = 1$ and $\lambda_n \neq 1$. Let λ_1 be arbitrary. Choose an integer k_1 such that $|A_{k_1}(\lambda_1)| < \varepsilon_1$ and a number $\eta_1 > 0$ such that $|A_{k_1}(\lambda) - 1| < \varepsilon_1$ whenever $|\lambda - 1| < \eta_1$. If λ_i, k_i, η_i are chosen for $1 \leq i \leq n$ then choose λ_{n+1} such that $|\lambda_{n+1} - 1| < \eta_i$ for all $i=1,\ldots,n$. Then choose $k_{n+1} > k_n$ such that $|A_{k_{n+1}}(\lambda_i)| < \varepsilon_{n+1}$ for $1 \leq i \leq n+1$ and choose $\eta_{n+1} > 0$ such that $|A_{k_{n+1}}(\lambda) - 1| < \varepsilon_{n+1}$ whenever $|\lambda - 1| < \eta_{n+1}$. Then the operator T and the sequence k_n satisfy the requirements.

Now Menchoff's result shows that on the L_2 space of the unit interval there is an orthonormal basis φ_n and a function $f \in L_2$ such that the sequence $Q_n f$ diverges a.e., with the previous definition of Q_n. Hence, if ε_n's are sufficiently small (e.g. satisfying $\sum_{n=1}^{\infty} \varepsilon_n < \infty$) and if T is the unitary operator given by the lemma above then $A_k(T)f$ can not converge a.e. . From these arguments it is also clear how to construct a contraction T of L_2 such that $\|Tf\| < \|f\|$ for each $f \in L_2$ for which (*) does not exist.

As mentioned before, the situation for a general contraction on an L_p space with $p \neq 2$ is not known. Here we would like to indicate that the arguments given above do not have obvious generalizations to other values of p. In fact, let φ_n be a basis for L_p, in the sense that each $f \in L_p$ has a unique strongly convergent expansion $f = \sum_{i=1}^{\infty} \alpha_i \varphi_i$, and let $Q_n : L_p \to L_p$ be defined as

$$Q_n\left(\sum_{i=1}^{\infty} \alpha_i \varphi_i\right) = \sum_{i=n+1}^{\infty} \alpha_i \varphi_i .$$

Then Q_n's must be contractions, if we hope to approximate them by $A_{n_k}(T)$. But, if $p \neq 2$ then a result of Ando [2] shows that Q_n's are conditional expections and, consequently, $Q_n f$ converges a.e. for each $f \in L_p$.

<u>References:</u>

1. AKCOGLU,M.A., A pointwise ergodic theorem in L_p spaces. Can.J.Math. 27, 1075-1082, (1975).
2. ANDO,T., Contractive projections in L_p-Spaces. Pac.J.Math.17,391-405, (1966).
3. BELLOW,A. (IONESCU-TULCEA,A.), Ergodic properties of isometries in L_p Spaces, Bull.A.M.S. 70, 366-371, (1964).
4. BIRKHOFF,G.D., Proof of the ergodic theorem, Proc.Nat.Acad.Sci.U.S.A. 17, 656-660, (1931).
5. BURKHOLDER,D.L., Semi Gaussian subspaces, Trans.A.M.S.104,123-131, (1962).
6. CHACON,R.V., A class of linear transformations, Proc.A.M.S.15, 560-564, (1964).

7. DUNFORD,N. and SCHWARTZ,J.T., Linear Operators, vol.1. Interscience Publishers New York, 1958.

8. HOPF,E., The general temporally discrete Markov processes, J.Rat. Mech.Anal. 3, 13-45, (1954).

9. MENCHOFF,D., Sur les sêries de fonctions orthogonales, Fund.Math. 4, 82-105, (1923).

Mustafa A. Akcoglu
Dept. of Mathematics
University of Toronto
Toronto M5S 1A1
Canada

GENERIC PROPERTIES OF MEASURE PRESERVING HOMEOMORPHISMS

by

Steve Alpern

Dept. of Mathematics
Yale University,
New Haven, Connecticut 06520
USA

1. <u>Introduction</u>. Let μ be the completion of a nonatomic, locally positive (i.e., on open sets) Borel measure on a compact connected metric space (X,d). It is well known [5 , p. 13] that under these assumptions the resulting measure space (X,Σ,μ) is a Lebesgue space.

We are concerned in this paper with the set $M = M(X, d,\mu)$ of μ-preserving homeomorphisms of X, and the question of whether the topology on X restricts the possible ergodic behavior of these homeomorphisms. For compact manifolds, it has been shown that ergodicity (Oxtoby and Ulam [13]) and weak mixing (Katok and Stepin [9], also [1]) are represented (in fact generically) in M.

One may ask what new properties (or isomorphism classes), if any, arise when one drops the continuity assumption and considers the larger class $G = G(X,\Sigma,\mu)$ of invertible μ-preserving transformations of the Lebesgue space (X,Σ,μ). For the two dimensional torus T^2, Lind and Thouvenot [10] have recently established that no new properties with finite entropy appear. Specifically, they show that every finite entropy ergodic automorphism in G is isomorphic to something in $M(T^2)$. In this paper we show that for spaces (X, d,μ) satisfying a condition we call NBD (for "norm-bounded density"), no new <u>generic</u> properties appear (when going from M to G). In a sense made precise in Theorem 3, we prove that any measure theoretic property generic in G with respect to the weak topology is generic in M with respect to the compact-open topology. This correspondence of generic properties was recently established [3] for spaces satisfying both the NBD <u>and</u> fixed point properties (in particular, for I^n), and it is only the removal of the latter hypothesis that is new in this paper. However, since NBD spaces include all compact manifolds of dimension n, $2 \le n \le \infty$ (where $n = \infty$ denotes Hilbert manifolds) [2], the topological generality of the result is significantly increased.

The principal technical advance of this paper is the following result on the conjugacy class (in G) of an arbitrary aperiodic automorphism.

Conjugacy Lemma: Let g and h be in G, with g aperiodic. Then for any $\varepsilon > 0$, there is an f in G satisfying

$$d(f^{-1}g\, f(x),\ h(x)) < \varepsilon \qquad \text{a.e.}\ .$$

The above result was recently established ([3], Topological Conjugacy Lemma) for the case where h is a homeomorphism with a fixed point. The present version is now an extension (rather than analog) of Halmos' Conjugacy Lemma [8, p. 77], since it has the same hypotheses and asserts the density of $\{f^{-1}\, g\, f\colon f \in G\}$ with respect to a finer topology (compact-open vs. weak).

The paper is organized as follows. In section 2 we prove a purely measure theoretic result (Theorem 1) which we had originally viewed as a generalization of Rohlin's Tower Theorem. At the conference, U. Krengel observed that Theorem 1 may also be seen as "almost" a special case of Theorem 1.1 of [6]. (It would be a special case if we restricted g to be ergodic rather than aperiodic.) However, rather than relying on the deep results of [6], we have decided to retain our original proof. In section 3, we show how Theorem 1 may be used to deduce the Conjugacy Lemma (and almost conversely). In section 4 we first consider homeomorphisms. We define the NBD condition, prove a property of NBD spaces (Theorem 2), and prove our main result (Theorem 3).

2. Generalized Rohlin Towers. The previous (restricted) proof of the Conjugacy Lemma used the following well known result.

Rohlin Tower Theorem (see [8], p. 71 for a proof): Let g in G be aperiodic. Then for every positive α and positive integer M, there is a (M-Rohlin) partition $S = \{S_i\}_{i=1}^{M}$ with $g\, S_i = S_{i+1}$ for $i = 1, \ldots, M-2$, and $\mu(S_M) < \alpha$. (In such a situation we define the "top" of the partition S, denoted T(S), by $T(S) = S_{M-1} \cup S_M$.)

It is now clear that the fixed point assumption in the Conjugacy Lemma of [3] was required because of the possibility that $\mu(g\, S_M \cap S_M) > 0$ when Rohlin's Theorem was appealed to. To generalize this result to partitions $\{S_i\}$

where $\mu(g\, S_i \cap S_j)$ is to be zero for certain pairs i,j specified in advance, we first must introduce some definitions.

An $n \times n$ matrix B consisting of 0's and 1's is said to be aperiodic if B^N has all positive entries (written $B^N > 0$) for some positive integer N. Two $n \times n$ matrices are called equivalent if their corresponding entries have the same sign. A probability distribution $(p_1, \ldots, p_n)$ is said to be consistent with an $n \times n$ 0-1 matrix B if it is invariant under some stochastic matrix (p_{ij}) equivalent to B. We can now state the main result of this section and two corollaries, the second of which shows why we consider these results as generalizations of the Tower Theorem.

Theorem 1: Let $(p_1, \ldots, p_n)$ be a consistent probability distribution for the $n \times n$ aperiodic 0-1 matrix B. Then for any aperiodic transformations g in G, there is a partition $P = \{P_i\}_{i=1}^n$ of X with $\mu(P_i) = p_i$, $i = 1, \ldots, n$, and $\mu(g\, P_i \cap P_j) = 0$ for i,j with $b_{ij} = 0$.

Corollary 1: Let (p_{ij}), $i, j = 1, \ldots, n$, be an aperiodic stochastic matrix (mixing Markov chain). Then for any aperiodic g in G there is a partition $P = \{P_i\}_{i=1}^n$ with $\mu(g\, P_i \cap P_j) = p_{ij}\ \mu(P_i)$

Corollary 2: For any $K \geq 2$, let $n_1, \ldots, n_K$ be relative prime integers and let $q_1, \ldots, q_K$ be positive numbers such that $n_1 q_1 + \ldots + n_K q_K = 1$. Then for any aperiodic g in G there exist sets Q_i, $i = 1, \ldots, K$ with $\mu(Q_i) = q_i$ and such that $X = \bigcup_{i=1}^{K} \bigcup_{j=0}^{n_i - 1} g^j(Q_i)$ is a partition (into K stacks of heights $n_1, \ldots, n_K$).

The proof of Theorem 1 is based on three lemmas, for which we need the following notation. Fix $g, n, B, (p_1, \ldots, p_n)$ satisfying the hypotheses of Theorem 1, and fix N with $B^N > 0$. All partitions considered (except Rohlin partitions) will have n elements, and for these we define a (complete) metric $\|Q-R\| = \sum_{i=1}^{n} \mu(Q_i \,\Delta\, R_i)$, where Δ denotes symmetric difference. For any partition R we define $W(R) = \{x \text{ in } X\colon x \in R_i \text{ and } g(x) \in R_j \text{ for } i,j \text{ with } b_{i,j} = 0\}$. In this notation, the last assertion of Theorem 1 is that $\mu(W(P)) = 0$.

It is useful to have a description of the set $D(B)$, consisting of all probability distributions consistent with B, as a subset of $\mathbb{R}^n$. We shall only need to know that $D(B)$ is the inner (or relative interior or algebraic interior) set of a convex polyhedron. To describe this set, let Γ denote the power set of $\{1, \ldots, n\}$ and define $\phi: \Gamma \longrightarrow \Gamma$ by $j \in \phi(A)$ if $b_{ij} = 1$ for some $i \in A$. Let Γ_1 be the subalgebra of Γ given by $A \in \Gamma_1$ if $b_{ij} = 1$ and $j \in \phi(A)$ imply $i \in A$. $D(B)$ can now be written (the proof is not hard) as the set of all probability distributions $y = (y_1, \ldots, y_n)$ satisfying

$$\sum_{i \in A} y_i = \sum_{j \in \phi(A)} y_j \quad \text{for } A \in \Gamma_1, \text{ and} \tag{2.1}$$

$$\sum_{i \in A} y_i < \sum_{j \in \phi(A)} y_j \quad \text{for } A \in \Gamma - \Gamma_1 . \tag{2.2}$$

At this point fix $\lambda > 0$ to be half the distance from $(p_1, \ldots, p_n)$ to the "boundary" of $D(B)$, that is, the set $\overline{D(B)} - D(B)$. For distances such as this in $\mathbb{R}^n$, we will always use the metric $|y^1 - y^2| = \sum_{i=1}^{n} |y_i^1 - y_i^2|$. For any partition R, let δR denote its distribution $(\mu R_1, \ldots, \mu R_n)$ and let $\delta(R/Y)$ denote its relative distribution on the subset Y, $1/\mu(Y) \cdot (\mu(R_1 \cap Y), \ldots, \mu(R_n \cap Y))$. We will also use the notation δV to denote an n-tuple (distribution) when no partition V has been mentioned. In particular we now define $\delta P = (p_1, \ldots, p_n)$, although a partition P with this distribution remains to be found. Finally, for any distribution δV, let $\delta^* V$ denote its orthogonal Euclidean projection onto the affine subspace containing $D(B)$, the subspace defined by (2.1).

<u>Lemma 1</u>: Let $S = \{S_i\}_{i=1}^M$ be a Rohlin partition for g, and let $(q_1, \ldots, q_n) \equiv \delta Q$ be in $D(B)$. Then there is a partition R with $\delta R = (q_1, \ldots, q_n)$ and $W(R) \subset T(S)$.

Proof: By definition of $D(B)$, $(q_1, \ldots, q_n)$ is invariant under a stochastic matrix p_{ij} equivalent to B. We define R recursively on $S_1, S_2, \ldots, S_{M-1}$ and then separately on S_M. On S_1 define R so that $\delta(R/S_1) = \delta Q$. If R has been defined on S_K, $K \le M-2$, define R on S_{K+1} by requiring $\delta(R/g(S_K \cap R_i)) = (p_{i1}, p_{i2}, \ldots, p_{in})$. This recursively defines R on $X - S_M$. On S_M define R so that $\delta(R/S_M) = \delta Q$. This definition ensures that $\delta R = \delta Q$, and the fact that $p_{ij} = 0$ whenever $b_{ij} = 0$ guarantees that $W(R) \subset T(S)$.

Lemma 2: Let $\beta > 0$ and let R be any partition of X. Then there is a Rohlin partition S with $\mu(T(S)) < \beta$ and a partition Q satisfying $W(Q) \subset T(S)$ and $\|Q-R\| < 2(N-1)\mu(W(R))$.

Proof: Rohlin's Theorem gives us the $S = \{S_i\}_{i=1}^M$ with $\mu(T(S)) < \beta$. The algorithm for defining Q consists of coding the R-columns of S. Consider a typical column whose base is the set of x in S_1 such that $g^j(x) \in R_{i_j}$ for $j = 0, \ldots, M-2$. We code the sequence $(i_0, \ldots, i_{M-2})$ into another sequence $(i_0', \ldots, i_{M-2}')$ satisfying

(2.3) $$b_{i_j', i_{j+1}'} = 1$$

for $j = 0, \ldots, M-3$. Let j_0 denote the least j with $b_{i_j, i_{j+1}} = 0$. Since $B^N > 0$ we may fill in the blanks in the sequence $(i_0, i_1, \ldots, i_{j_0}, -, -, \ldots, -, i_{j_0+n})$ so that the resulting sequence will have no transitions corresponding to 0's of B. Apply the same process beginning at the first place j_1 $(j_1 \ge j_0+n)$ where another "illegal" transition occurs. When the coded sequence $(i_1', \ldots, i_{M-1}')$ satisfying (2.3) is found, define Q by setting $g^j(x) \in Q_{i_j'}$ for $j = 0, \ldots, M-2$, and $x \in S_1$. The fact that i_j' satisfies (2.3) ensures that $W(Q) \subset T(S)$. The fact that in going from $\{i_j\}$ to $\{i_j'\}$ every illegal transition resulted in no more than N-1 changes guarantees that $\|Q-R\| \le 2(N-1)\ \mu(W(R))$.

Unfortunately, δQ may be even further from $(p_1, \ldots, p_n) \equiv \delta P$ than δR is. To correct this, in the next lemma we "average" Q with another partition L so that the distribution $\frac{1}{1+\theta}\ \delta Q + \frac{\theta}{1+\theta}\ \delta L$ is near δP, and that its projection onto D(B) is δP.

Lemma 3: Let $S = \{S_i\}_{i=1}^M$ be a Rohlin partition. Let Q be a partition with $|\delta^* Q - \delta P| < \lambda$ and $W(Q) \subset T(S)$. Then there is a partition U of X with

$W(U) \subset T(S)$, $\delta^* U = \delta P$ and $\|U-Q\| < (2/\lambda) \, |\delta^* Q - \delta P|$.

<u>Proof</u>: (At first reading it is helpful to assume that $\dim D(B) = n-1$, so that $\delta^* = \delta$ for all partitions.) First observe that $\delta^*(Q) \in D(B)$, because $|\delta^*(Q) - \delta P| < \lambda$. Define θ by $|\delta^* Q - \delta P| = \theta \lambda$ and define δL by

$$\delta L = \delta P - 1/\theta \, (\delta^* Q - \delta P).$$

A simple calculation shows that

$$\delta P = \frac{1}{1+\theta} \delta^* Q + \frac{\theta}{1+\theta} \delta L .$$

By convexity, this implies that δL lies in the affine subspace containing $D(B)$, defined by equations (2.1). Furthermore $|\delta L - \delta P| = \lambda$, so that $\delta L \in D(B)$.

Let C_k, $k = 1, \ldots, K$, be the columns of the Rohlin Tower $\{S_i\}_{i=1}^{M-1}$ defined by partitioning S_1 into $Q - (M - 1)$ names. Divide column C_k into two columns C_k^1 and C_k^2 with $\mu(C_k^2) = \theta\mu(C_k^1)$. Similarly divide the exceptional set S_M into two sets C_0^1 and C_0^2 so that $\delta(Q/C_0^1) = \delta(Q/C_0^2)$. Finally, partition $X = Y^1 \cup Y^2$ by defining $Y^i = \bigcup_{k=0}^{K} C_k^i$, $i = 1,2$. Observe that $\mu(Y^1) = \frac{1}{1+\theta}$ and $\mu(Y^2) = \frac{\theta}{1+\theta}$ and that $\delta(Q/Y_1) = \delta(Q/Y_2) = \delta Q$. We define the partition U separately on Y^1 and Y^2. On Y^1, set $U = Q$. Define U on Y^2 using Lemma 1 (or more technically, its proof) so that $\delta(U/Y_2) = \delta L$. Since $Q = U$ on Y^1, and $\mu(Y^1) = 1/1+\theta$, we have

$$\|U-Q\| \le \frac{2\theta}{1+\theta} < 2\,\theta = (2/\lambda) \, |\delta^* Q - \delta P|.$$

Furthermore, $\delta U = \mu(Y^1)\delta(U/Y^1) + \mu(Y^2)\, \delta(U/Y^2)$

$$= \frac{1}{1+\theta} \delta Q + \frac{\theta}{1+\theta} \delta L .$$

Taking projections, we have

$$\delta^* U = \frac{1}{1+\theta} \delta^* Q + \frac{\theta}{1+\theta} \delta^* L = \delta P.$$

<u>Proof of Theorem 1:</u> Choose β_k, $k = 1,2, \ldots$ with $\beta_k < \lambda/2N$ and

$\sum_{k=1}^{\infty} \beta_k < \infty$. Using Lemma 1 define a partition R^1 with $\delta R^1 = \delta P$ and $\mu(W(R^1)) < \beta_1$. We now recursively construct a sequence $\{R^k\}_{k=2}^{\infty}$ satisfying

(2.3) $$\mu(W(R^k)) < \beta_k,$$

(2.4) $$\delta^* R^k = \delta P, \text{ and}$$

(2.5) $$\|R^k - R^{k-1}\| < C\ \beta_{k-1},$$

where $C = 4N/\lambda + 2N$. Clearly R^1 satisfies (2.3) and (2.4). Suppose that partitions R^k satisfying (2.3) and (2.4) have been found for $k < j$. We then construct R^j satisfying (2.3), (2.4) and (2.5) as follows. Apply Lemma 2 with $R = R^{j-1}$ and $\beta = \beta_j$ to find a Rohlin partition S with $\mu(T(S)) < \beta_j$ and a partition Q satisfying $W(Q) \subset T(S)$ and $\|Q-R^{j-1}\| < 2(N-1)\mu(W(R^{j-1})) < 2(N-1)\ \beta_{j-1} < \lambda$. Since $\delta^* R^{j-1} = \delta P$, we have

$$|\delta^* Q - \delta P| \leq |\delta Q - \delta R^{j-1}| < \|Q-R^{j-1}\| < \lambda \quad .$$

Consequently Q satisfies the hypotheses of Lemma 3 with respect to S, and therefore there is a partition U with $W(U) \subset T(S)$, $\delta^* U = \delta P$, and $\|U-Q\| < (2/\lambda)\ |\delta^* Q - \delta P| < 4N\ \beta_{j-1}/\lambda$. Setting $R^j = U$, it is immediate that R^j satisfies (2.3) and (2.4). To check (2.5) we estimate

$$\begin{aligned}\|R^j - R^{j-1}\| &\leq \|R^j - Q\| + \|Q - R^{j-1}\| \\ &\leq (4N/\lambda)\ \beta_{j-1} + 2N\ \beta_{j-1} \\ &\leq C\ \beta_{j-1}\ .\end{aligned}$$

We now observe that (2.5) implies that the R^k form a Cauchy sequence with respect to the partition metric $\|\ -\ \|$. Since this metric is complete there is a partition $\hat{P}$ with $\|R^k - \hat{P}\| \longrightarrow 0$. Since both $\mu(W(\cdot))$ and $\delta^*(\cdot)$ are continuous with respect to the partition metric, it follows that $\mu(W(\hat{P})) = 0$ and $\delta^* \hat{P} = \delta P$, where δP still is shorthand for the distribution $(p_1, \ldots, p_n)$ and is not necessarily the distribution of P, which is not yet defined. But since $\mu(W(\hat{P})) = 0$ it follows that $\hat{P}$ belongs to $D(B)$ by defining

$p_{ij} = \mu(g\,\hat{P}_i \cap \hat{P}_j)/\mu(\hat{P}_i)$. For elements of $D(B)$, $\delta^* = \delta$, so in fact $\delta\,\hat{P} = \delta P = (p_1, \ldots, p_n)$. The partition $P = \hat{P}$ satisfies all the requirements of the Theorem.

3. <u>Conjugacy Lemma.</u> In this section we prove the Conjugacy Lemma using Theorem 1. We also show that for "non-critical" matrices B, we may prove Theorem 1 using the Conjugacy Lemma. Thus, Theorem 1 and the Conjugacy Lemma are in the same relationship as Rohlin's Tower Theorem and Halmos' Conjugacy Lemma.

<u>Conjugacy Lemma</u>: Let g and h be in G, with g aperiodic. Then for any $\epsilon > 0$ there is an f in G satisfying

$$d(f^{-1}g\,f(x),\, h(x)) < \epsilon \quad , \quad \text{a.e.}$$

<u>Proof</u>: Let $Q = \{Q_i\}_{i=1}^n$ be a measurable partition of X satisfying both $d(h\,Q_i) < \epsilon/2$ and $d(Q_i) < \epsilon/2$ for $i = 1, \ldots, n$ where d in this context denotes set diameter. Define an $n \times n$ $0 - 1$ matrix B by $b_{ij} = 1$ if $\overline{h\,Q_i} \cap \overline{Q_j} \neq \emptyset$ (bar denotes closure). We claim that B is aperiodic and that $\delta\,Q$ belongs to $D(B)$. To see this let Γ and ϕ be as defined prior to Lemma 1, and observe that

$$\sum_{i \in A} \mu Q_i \leq \sum_{j \in \phi(A)} \mu\,Q_j \quad \text{for} \quad A \in \Gamma \;. \tag{3.1}$$

Suppose equality holds in (3.1) for some $A \in \Gamma$ other than $\emptyset$ and $\{1, \ldots, n\}$. Then the set $Y = \left(\bigcup_{i \in A} \overline{h\,Q_i}\right) \cup \left(\bigcup_{j \in \phi(A)} \overline{Q_j}\right)$ is the disjoint union of two nonempty closed sets and $\mu(Y) = 1$. It follows that $X-Y$ is an open set with measure 0 and therefore empty (by assumption on μ), or $X = Y$. But then X is not connected, contradicting the assumption on X. Therefore strict inequality holds in (3.1) for all nonempty proper subsets A. This means, according to (2.1) and (2.2), that $\delta\,Q \in D(B)$, and that $\Gamma_1 = \{\emptyset, \{1, \ldots, n\}\}$. Choose an integer N sufficiently large so that $1/N$ is less than the minimum difference between the left and right hand sides of (3.1) for $A \in \Gamma - \Gamma_1$. Repeated application of (3.1) shows that $\sum_{j \in \phi^N(A)} \mu(Q_j) = 1$ for all A in Γ.

Consequently $\phi^N(A) = \{1, \ldots, n\}$ for all A in Γ, or $B^N > 0$, and B is aperiodic.

Now apply Theorem 1 to find a partition $P = \{P_i\}_{i=1}^n$ of X with $\mu(P_i) = \mu(Q_i)$, $i = 1, \ldots, n$, and $\mu(g\,P_i \cap P_j) > 0$ only if $\overline{h\,Q_i} \cap \overline{Q_j} \neq \emptyset$. Let f be any transformation in G satisfying $f(Q_i) = P_i$, $i = 1, \ldots, n$ (see [8, p. 74]). Fix $x \in X$ and suppose $x \in Q_i$. Then almost surely $f(x) \in P_i$ and $g\,f(x) \in P_j$ for some j with $\overline{h\,Q_i} \cap \overline{Q_j} \neq \emptyset$, and $f^{-1}g\,f(x) \in Q_j$. Also, $h(x) \in h(Q_i)$, so

$$\begin{aligned} d(f^{-1}\,g\,f(x), h(x)) &\leq d(Q_j \cup h\,Q_i) \\ &\leq d(Q_j) + d(h\,Q_i) \\ &\leq \epsilon/2 + \epsilon/2 = \epsilon\,. \end{aligned}$$

We now show that Theorem 1 may be proved via the Conjugacy Lemma in the case where the subalgebra Γ_1 of Γ corresponding to B is the trivial subalgebra. Such a B will be called "non-critical". Observe that non-critical implies aperiodic.

<u>Proposition</u>: Let B be a non-critical $n \times n$ $0-1$ matrix and let $(p_1,\ldots,p_n)$ belong to $D(B)$. Then for any aperiodic g in G there is a partition $P = \{P_i\}_{i=1}^n$ of X with $\delta\,P = (p_1, \ldots, p_n)$ and $\mu(g\,P_i \cap P_j) = 0$ if $b_{ij} = 0$.

<u>Proof</u>: We define (X,μ, d) as follows: Let ϕ and Γ correspond to B. Let X be an embedding into $\mathbb{R}^3$ of the combinatorial graph with n vertices (labeled 1, 2, ..., n) and with an edge between i and j if there is a k with $b_{ki} = b_{kj} = 1$. Since B is non-critical, it follows that this graph is connected and consequently that X is connected as a subset of $\mathbb{R}^3$ with the Euclidean metric d. When i is adjacent to j, let α_{ij} be an interior point of the line segment from i to j, and let Q_i be the union of the closed intervals from i to α_{ij} for all j adjacent to i. The Q_i, $i = 1, \ldots, n$ are compact subsets of X which intersect pairwise in at most one point (α_{ij}). Let μ be any locally positive nonatomic measure on X with $\mu(Q_i) = p_i$, $i = 1, \ldots, n$. We

now find $h \in G(X,\mu)$ and $\epsilon > 0$ so that if $d(f^{-1} g f(x),h(x)) < \epsilon$, a.e., then the sets $P_i = f(Q_i)$ will have the required properties.

Let $\beta > 0$ be chosen so that the two sides of the inequalities (2.2) differ by more than α when $(y_1, \ldots, y_n) = (p_1, \ldots, p_n)$. Choose $\epsilon > 0$ such that $\mu(N_\epsilon(\bigcup_{i,j} \alpha_{ij})) < \beta$, where N_ϵ denotes the Euclidean ϵ-neighborhood of a subset of X.

We now use a "marriage" argument. We say that each x in Q_i (or simply Q_i) "knows" all points in the set

$$\bigcup_{j \in \phi\{i\}} Q_j - \bigcup_{\substack{j \in \phi\{i\} \\ k \notin \phi\{i\}}} N_\epsilon \{\alpha_{j,k}\} .$$

It follows that the set Q_i, $i \in A \in \Gamma - \Gamma_1$, together "know" all points in the set

$$\bigcup_{j \in \phi(A)} Q_j - \bigcup_{\substack{j \in \phi(A) \\ k \notin \phi(A)}} N_\epsilon \{\alpha_{j,k}\} .$$

But since the measure of this set, by construction, exceeds $\sum_{i \in A} \mu(Q_i)$, the marriage condition is fulfilled. It follows by an easy application of the combinatorial marriage theorem, or by the "measure marriage theorem" that there is an $h \in G(X,\mu)$ with x "knows" h(x) a.e. . If $x \in Q_i$ and $d(f^{-1} g f(x),h(x)) < \epsilon$ then $f^{-1} g f(x)$ belongs to a Q_j with $j \in \phi\{i\}$, or $b_{ij} = 1$. Consequently the sets defined by $P_i = Q_i$ have the required properties.

4. <u>NBD Spaces</u>. In this section we discuss generic ergodic properties of μ-preserving homeomorphisms of NBD spaces (X, d, μ). First, we define two topologies, the weak and compact-open (called "norm" in [3]), on $G = G(X, \Sigma, d, \mu)$. Recall that G is the group of invertible μ -preserving transformations on X. The weak topology is defined by the subbasic family of neighborhoods $N = N(f, Y, \alpha) = \{g \in G: \mu(gY \Delta fY) < \alpha\}$, where $f \in G$, $Y \in \Sigma$ and $\alpha > 0$. The compact-open topology depends on the metric d and is

given by the metric $\rho(f,g) = \text{ess sup } d(f(x),g(x))$. The topology induced by ρ on the subgroup $M = M(X, d, \mu)$ of G consisting of homeomorphisms is simply the topology of uniform convergence. For f in G we define the norm of f, denoted $\|f\|$, by $\|f\| = \rho(f, \text{identity})$. We can now define the NBD condition.

Definition: A metric measure space (X, d, μ) has the NBD ("norm-bounded density") property if for all $\epsilon > 0$ there is a $\delta > 0$ such that for any $g \in G(X, d, \mu)$ with $\|g\| < \delta$, and weak neighborhood N of g, there is a homeomorphism $h \in M(X, d, \mu)$ with $\|h\| < \epsilon$ and $h \in N$.

In [2] it is shown that if μ is product Lebesgue measure and (X,d) is the Euclidean n-cube, $n \geq 2$, or the Hilbert cube, then (X, d, μ) satisfies the NBD condition with $\delta = \epsilon$. It is also indicated there how the same proof applies when (X,d) is obtained from the cube by making identifications on the boundary. We observe that if (X, d, μ) is NBD, then so is $(X, d, \mu h)$ for any homeomorphism h of (X,d). Since for the cube [13] or Hilbert cube [12] any locally positive nonatomic measure can be represented as mh where m is Lebesgue measure, the above spaces (X,d) are NBD regardless of the measure.

Actually the spaces mentioned above satisfy a stronger condition where the weak neighborhood N is replaced (in the NBD definition) by a uniform topology (see [8] for definition) neighborhood. This approach was initiated independently by Oxtoby [11] and White [14] and followed by Edwards and the author [4].

Theorem 2: Let (X, d, μ) be an NBD space and let $G = G(X, d, \mu)$ and $M = M(X, d, \mu)$. Let V be a G_δ subset of G in the weak topology and assume that the compact-open topology closure of V contains M. Then $V \cap M$ is a dense G_δ subset of M in the compact-open topology.

Proof: With minor modifications, the proof follows that of Corollary 4.1 of [2]. The Baire Category Theorem is used in this proof.

Theorem 3: Let (X, d, μ) be an NBD space. If V is any conjugate-invariant subset of G which is dense G_δ in the weak topology, then $V \cap M$ is a dense G_δ subset of M in the compact-open topology.

Proof: Since V is dense G_δ in G it must contain an aperiodic transformation g. Consequently the conjugat-invariant set V must also contain the conjugacy class, $\{f^{-1} g f: f \in G\}$, of g in G. The Conjugacy Lemma now implies that V is dense in G in the compact-open topology. Therefore V satisfies the conditions of Theorem 2 and so $V \cap M$ is a dense G_δ subset of M in the compact-open topology.

BIBLIOGRAPHY

[1] S. Alpern, New proofs that weak mixing is generic, Invent. Math. 32 (1976), 263-278.

[2] S. Alpern, Approximation to and by measure preserving homeomorphisms, Journal of the London Maths. Soc., to appear.

[3] S. Alpern, A topological analog of Halmos' Conjugacy Lemma, Invent. Math., 48, (1978), 1-6.

[4] S. Alpern and R. D. Edwards, Lusin's Theorem for measure preserving homeomorphisms, to appear, in: Mathematika.

[5] M. Denker, C. Grillenberger and K. Sigmund, Ergodic Theory on Compact Spaces, Lecture Notes in Mathematics, Vol. 527, Springer, Berlin (1973), 206-217.

[6] C. Grillenberger and U. Krengel, On marginal distributions and isomorphisms of stationary processes, Math. Zeit. 149 (1976), 131-154.

[7] P. Halmos, In general, a measure preserving transformation is mixing. Ann. of Math. 45 (1944), 786-792.

[8] P. Halmos, Lectures on ergodic theory, Chelsea, New York, 1956 .

[9] A. Katok and A. Stepin, Metric properties of measure preserving homeomorphisms, Uspekhi Mat. Nauk. 25:2 (1970), 193-220. (Russian Math. Surveys 25 (1970), 191-220.)

[10] D. Lind and J. Thouvenot, Measure preserving homeomorphisms of the torus represent all finite entropy ergodic transformations, Math. Systems Theory II (1978), 275-282.

[11] J. Oxtoby, Approximation by measure preserving homeomorphisms, Recent Advances in Topological Dynamics; Lecture Notes in Mathematics Vol. 318, Springer, Berlin (1973), 206-217.

[12] J. Oxtoby and V. Praaad, Homeomorphic measures in the Hilbert cube, Pacific J. Math., to appear.

[13] J. Oxtoby and S. Ulam, Measure-preserving homeomorphisms and metrical transitivity, Ann. of Math. (2) -2 (1941), 874-920.

[14] H. E. White, Jr., The approximation of one-one measurable transformations by measure preserving homeomorphisms, Proc. Amer. Math. Soc. 44 (1974), 391-394.

ON DISJOINTNESS IN TOPOLOGICAL DYNAMICS AND ERGODIC THEORY

by

Joseph Auslander

In [5], Furstenberg introduced the notions of disjointness of flows and processes. Recall that two processes (measure preserving transformations on probability spaces) (X,φ,μ) and (Y,ψ,ν) are disjoint if the only $\varphi \times \psi$ invariant probability measure on $X \times Y$ which projects to μ and ν is the product measure. Two flows (homeomorphisms of compact metric spaces) (X,φ) and (Y,ψ) are (topologically) disjoint if the only closed invariant subset of $X \times Y$ which projects to X and Y is the product space $X \times Y$. If (X,φ) and (X,ψ) are minimal then disjointness is equivalent to minimality of the product system $(X \times Y, \varphi \times \psi)$.

In both the measure theoretic and topological categories, we use a perpendicular sign, $\perp$, to denote disjointness.

Suppose (X,φ) and (Y,ψ) are minimal flows, equipped with invariant Borel probability measures μ and ν respectively. It is natural to consider the relation between the two kinds of disjointness. In general, neither of measure theoretic nor topological disjointness implies the other. In fact, it is possible to construct a minimal flow which supports two disjoint ergodic measures (Y. Katznelson, personal communication). An example of topologically disjoint minimal flows which support non-disjoint ergodic measures will be given after the proof of Theorem 2.

Our first theorem gives two sufficient conditions for measure theoretic disjointness to imply topological disjointness.

Theorem 1. Let (X,φ) and (Y,ψ) be minimal flows, and let μ and ν be ergodic invariant Borel measures on X and Y respectively, such that the processes (X,φ,μ) and (Y,ψ,ν) are disjoint. Suppose (X,φ) satisfies one of the following three conditions: (i) point distal (ii) the proximal relation is an equivalence relation (iii) unique ergodicity. Then (X,φ) and (Y,ψ) are topologically disjoint.

That unique ergodicity is sufficient is proved in [2] (Theorem 10). The sufficiency of the other conditions is a consequence of a theorem on transformation groups with compact Hausdorff phase spaces. Before stating this theorem, we review several dynamical notions. The transformation group (X,T) is called point transitive if there is a point x_0 with a dense orbit (x_0 is called a transitive point). The points x and y are proximal if there is a net $\{t_n\}$ in T and $z \in X$ such that $xt_n \to z$ and $yt_n \to z$. We write $P(X)$ for the proximal relation in X. If (X,T) and (Y,T) are transformation groups the extension (homomorphism) $\pi: X \to Y$ is proximal if, for every $y \in Y$ and $x_1,x_2 \in \pi^{-1}(y)$, x_1 and x_2 are proximal, and π is highly proximal, if, for each $y \in Y$, there is a net $\{t_n\}$ in T and an $x \in X$ such that $\pi^{-1}(y)t_n \to \{x\}$ (in the Hausdorff topology on X). If X and Y are minimal π is highly proximal if and only if every non-empty open set in X contains a fiber $\pi^{-1}(y)$. When X and Y are metric spaces highly proximal is equivalent to almost one to one (some fiber is a singleton), [1].

The minimal flow (X,T) is called HPI ("highly proximal isometric") if there is a highly proximal extension $X^{\sim}$ of X such that $X^{\sim}$ is obtained from the trivial (one point) flow by a succession (in general transfinite) of highly proximal and almost periodic extensions. That is, there is an ordinal number η and a "tower" of flows and homomorphisms

$$X_0 = \tilde{X_0} \leftarrow X_1 \leftarrow \tilde{X_1} \leftarrow X_2 \leftarrow \tilde{X_2} \leftarrow \cdots \leftarrow X_\alpha \leftarrow \tilde{X_\alpha} \leftarrow \cdots \leftarrow X_\eta \leftarrow \tilde{X_\eta}$$

where $X_0 = \tilde{X_0} = \{1\}$, $\tilde{X_\eta} = \tilde{X}$, the extensions $\tilde{X_\alpha} \to X_\alpha$ and $X_{\alpha+1} \to \tilde{X_\alpha}$ are, respectively, highly proximal and almost periodic, and, if α is a limit ordinal $X_\alpha = \lim_{\zeta<\alpha} X_\zeta$. (By the "relativized" Furstenberg structure theorem, [3], "almost periodic" may be replaced by "distal" throughout in this definition, at least in the presence of a separability hypothesis.) As we have remarked, highly proximal is equivalent to almost one-to-one if the spaces are metric. Therefore, by the Veech structure theorem ([7], [4]) HPI is equivalent to point distal (some point is distal to all other points) in the metric case.

<u>Theorem 2</u>. Let (X,T) and (Y,T) be minimal transformation groups such that $(X \times Y, T)$ is point transitive. Suppose either (i) (X,T) is HPI or (ii) T is abelian, and the proximal relation X is an equivalence relation. Then $(X,T) \perp (Y,T)$.

Before proving Theorem 2, we show how it implies Theorem 1. Since μ and ν are ergodic, and the processes (X,φ,μ) and (Y,ψ,ν) are disjoint, $\mu \times \nu$ is ergodic ([2]), and (since (X,φ) and (Y,ψ) are minimal) $\mu \times \nu$ is positive on open sets. Thus there is a dense orbit in $X \times Y$, that is, $(X \times Y, \varphi \times \psi)$ is point transitive. Since point distal minimal flows are HPI, Theorem 2 obviously applies.

The proof of Theorem 2 depends upon 3 lemmas:

<u>Lemma 1</u>. Let (X,T), (X',T), (Y,T) be minimal. Suppose $(X \times Y, T)$ is point transitive, and X' is a highly proximal extension of X. Then $(X' \times Y, T)$ is point transitive.

<u>Lemma 2</u>. Let (X,T) and (Y,T) be minimal, and suppose $P(X)$ is an equivalence relation. Then there is a unique minimal set in $X \times Y$.

Lemma 3. Let (X,T) and (Y,T) be minimal, with T abelian. Suppose there is a unique minimal set in $X \times Y$. Then $(X,T) \perp (Y,T)$.

Proofs: Lemma 1: Let (x_0,y_0) be a transitive point in $X \times Y$. Let $\pi: X' \to X$ be a highly proximal extension, and let $x_0' \in \pi^{-1}(x_0)$. We show that (x_0',y_0) is a transitive point in $X' \times Y$. Let U' and V be open in X' and Y respectively, and let $x \in X$ with $\pi^{-1}(x) \subset U'$. Let $y \in V$ and let $\{t_n\}$ be a net in T such that $(x_0,y_0)t_n \to (x,y)$. Let (a subnet of) $x_0't_n \to x'$. Then $x' \in \pi^{-1}(x) \subset U'$ and $(x_0',y_0)t_n = (x_0't_n,y_0t_n) \in U' \times V$ $(n \geq n_0)$. This completes the proof.

Lemma 2: Let $z = (x_0,y_0)$ be a transitive point, and suppose M_1 and M_2 are minimal subsets of $X \times Y$. Let I_1 and I_2 be minimal right ideals in $E(X \times Y)$, the enveloping semigroup of $X \times Y$, such that $z_0I_1 = M_1$, $z_0I_2 = M_2$, and let u_1 and u_2 be idempotents in I_1 and I_2 respectively such that $y_0u_1 = y_0$, $y_0u_2 = y_0$. Now $(x_0u_1,y_0) = z_0u_1 \in M_1$, $(x_0u_2,y_0) = z_0u_2 \in M_2$, $(x_0,x_0u_1) \in P(X)$, $(x_0,x_0u_2) \in P(X)$, so $(x_0u_1,x_0u_2) \in P(X)^2 = P(X)$. It follows easily that $(z_0u_1,z_0u_2) \in P(X \times Y)$. But $z_0u_1 \in M_1$, $z_0u_2 \in M_2$, and points in distinct minimal sets cannot be proximal. Hence $M_1 = M_2$.

Lemma 3. Since T is abelian, the almost periodic points are dense in $X \times Y$ (for, if (x,y) is almost periodic, then so is (xt,ys), for $t,s \in T$). Now all almost periodic points are in M, the unique minimal set. It follows that $M = X \times Y$.

Now we prove Theorem 2. If T is abelian and $P(X)$ is an equivalence relation, the proof follows immediately from Lemmas 2 and 3. Now suppose (X,T) is HPI. Then there is a highly proximal extension $X^\sim$ of X which has an "HPI tower" as in the definition. By Lemma 1, $(X^\sim \times Y, T)$ is point transitive. Thus, for all $\alpha \leq \eta$, $(X_\alpha \times Y, T)$ and $(X_\alpha^\sim \times Y, T)$ are point transitive. Clearly, $X_0 \times Y = X_0^\sim \times Y = Y$ is minimal. Let $\alpha \leq \eta$, and suppose $X_\beta \perp Y$ and $X_\beta^\sim \perp Y$ for all

$\beta < \alpha$. If α is a limit ordinal, $X_\alpha \perp Y$, since disjointness is preserved under passage to limits. Suppose $\alpha = \gamma + 1$. Since $\tilde{X_\gamma} \perp Y$ and the extension $X_\alpha \to \tilde{X_\gamma}$ is almost periodic $(X_\alpha \times Y, T)$ is pointwise almost periodic. Since $(X_\alpha \times Y, T)$ is also point transitive, it must be minimal; that is $X_\alpha \perp Y$. Now ([1], Theorem I.2) disjointness is always preserved under highly proximal extensions. Thus $\tilde{X_\alpha} \perp Y$, and the proof is completed.

To conclude, we indicate why topological disjointness does not imply measure theoretic disjointness. P. Julius ([6]) has constructed an example of a point distal minimal flow (X,φ) with positive topological entropy. Let (Y,ψ) be a (topologically) weak mixing minimal flow with positive topological entropy. Then there are ergodic measures μ and ν such that the entropies $h(\varphi,\mu)$ and $h(\psi,\nu)$ are positive. Thus the processes (X,φ,μ) and (Y,ψ,ν) cannot be disjoint, [5]. On the other hand, it follows from [1], Theorem II.1, that any HPI minimal flow (hence any point distal one) is topologically disjoint from any minimal flow which is disjoint from all distal minimal flows. But the latter class coincides with the weakly mixing minimal flows. Thus (X,φ) and (Y,ψ) are topologically disjoint.

REFERENCES

1. J. Auslander and S. Glasner, Distal and highly proximal extensions of minimal flows, Indiana University Math. Journal 26(1977), 731-749.

2. J. Auslander and Y.N. Dowker, On disjointness of dynamical systems, to appear, Proceedings of Cambridge Philosophical Society.

3. R. Ellis, Lectures on Topological Dynamics, W.A. Benjamin, N.Y., 1969.

4. R. Ellis, The Veech structure theorem, Trans. Amer. Math. Soc. 186 (1973), 203-218.

5. H. Furstenberg, Disjointness in ergodic theory, minimal sets, and a problem in diophantine approximation, Math. Systems Theory, 1 (1967), 1-49.

6. P. Julius, A point distal homeomorphism with positive topological entropy, preprint.

7. W. Veech, Point-distal flows, Amer. J. Math. 92(1970), 205-242.

UNIVERSITY OF MARYLAND
DEPARTMENT OF MATHEMATICS
COLLEGE PARK, MARYLAND

Reparametrization of probability-preserving n-flows

by

J. Feldman

§1. Introduction. By an n-flow ϕ on the probability space (X,μ), we shall always mean a free, measure-preserving, ergodic action of $\mathbb{R}^n$ on (X,μ). By a reparametrizing map τ for ϕ we mean a jointly measurable $\tau : X \times \mathbb{R}^n \to \mathbb{R}^n$ such that each $\tau(x,\cdot)$ is a homeomorphism: $\mathbb{R}^n \to \mathbb{R}^n$ and such that the function $(x,v) \to \phi_{\tau(x,v)}(x)$ is again an n-flow on (X,μ_τ), where μ_τ is a certain probability measure equivalent to μ; ϕ_τ is then called a reparametrization of ϕ. If $T : (X, \mu_\tau) \to (Y,\nu)$ is a measure-isomorphism and $\psi_v = T\phi_{\tau(\cdot,v)}T^{-1}$, then T takes the measure class of μ to that of ν and sends orbits of ϕ homeomorphically to orbits of ψ: we call T a homeomorphic orbit equivalence between ϕ and ψ. Conversely, if T is a homeomorphic orbit equivalence linking ϕ and ψ, then $\psi_v T$ has the form $T\phi_{\tau(\cdot,v)}$, where τ is a reparametrizing map, and then $\psi = T\phi_\tau T^{-1}$. Thus, we may ask the question "when is ψ homeomorphically orbit-equivalent to ϕ?" or, what is the same question, "when is ψ isomorphic to a reparametrization of ϕ?" In the last three years a rich theory has developed around this question. In the following pages we shall describe the basic parts of this theory. For brevity, instead of "homeomorphic orbit-equivalence" we just say equivalence.

The author wishes to thank the Miller Institute for Basic Research in Science and the U.S.National Science Foundation (Grants # MCS 75-05576 and MCS 06718) for their support of this work.

§2. The case n = 1. For 1-flows, which - as is customary - we will just call flows, the question of homeomorphic orbit equivalence was discussed by S.Kakutani in 1942 [K]. At that time, it appeared possible that all flows were equivalent. Kakutani converted the problem into a problem about transformations (by which we shall always mean ergodic, aperiodic, measure-preserving, invertible transformations on probability spaces), by means of the theorem of W.Ambrose on flows built under functions. This theorem we will now recall.

Let T be a transformation on (X_o,μ_o) and $f : X_o \to \mathbb{R}^+$, integrable. Define a flow $\phi = \phi_T^f$, the flow built over T and under f, on the space $\{(x_o,s) : 0 \le s < f(x_o)\}$ with measure $d\mu = (\int f\, d\mu_o)^{-1} d\mu_o ds$, by the rule $\phi_t(x_o,s) = (T^n x_o, s+r)$. Here n is the smallest integer with $s + t < f(x_o) + \ldots + f(T^n x_o)$, and $r = s + t - (f(x_o) + \ldots + f(T^{n-1} x_o))$. What Ambrose showed in [Am] was that any flow has such a representation.

It is fairly straightforward to show that two flows are equivalent if and only if they may be written as flows over the same transformation (but usually under different functions, of course). Thus, the classification of flows up to equivalence is the same as the classification of transformations under the equivalence relation which says: S and T are equivalent if isomorphic flows may be build over them. Kakutani pointed out in [K] that this may be described more directly, via the notion of induced transformation, as follows.

If T is a transformation on (X,μ), then for any subset $A \subset X$ with $\mu(A) > 0$, a transformation T_A on (A,μ_A) may be defined, where $d\mu_A = 1/\mu(A)\ d\mu$, by setting $T_A x = T^{n(x)}$, where $n(x) = \inf\{n > 0 : T^n x \in A\}$. T_A is said to be induced on A by T . Then Kakutani's observation was that S and T can be bases for isomorphic flows if and only if they have isomorphic induced transformations. This notion of equivalence we have called Kakutani equivalence (in [F1]), while the Russians have called it monotone equivalence. Then the question remained: are all transformations Kakutani-equivalent?

The discovery in 1957 by Kolmogorov and Sinai of entropy quickly led to a negative answer. More precisely: Abramov, in [A], showed that $h(T) = (\int f d\mu) h(\phi_Y^f) = \mu(A)\ h(T_A)$. Thus, there are at least three inequivalent classes of transformations or of flows: entropy zero, entropy positive but finite, and entropy infinite. Here the matter rested for a number of years.

In 1975, in work of A.Katok, D.Ornstein, D.Rudolph, E.Satayev, B.Weiss, and the present author, some important developments took place. Briefly: within each of the three entropy classes there was found a single "simplest" equivalence class, the "loosely Bernoulli" (LB) transformations: simplest in the sense that any factor of an LB transformation is LB, and that given any LB transformation S and any transformation T of large enough entropy class, then $\exists$ T' equivalent to T with S a factor of T'. Examples of LB transformations of zero entropy are transformations with purely discrete spectrum; for positive entropy: Bernoulli shifts. Finally, returning to our question: within each entropy class there exist non LB transformations; in fact, there exist uncountably many inequivalent transformations! This work may be found in [F1], [Ka1], [Ka2], [R1], [S], [W].

Passing to flows: we say a flow is LB if it may be written as a flow built over an LB transformation. Another equivalent definition (although this is a very deep fact) is that ϕ is an LB flow if some (equivalently, any) ergodic ϕ_t is an LB transformation. Examples of LB flows of zero entropy are given by flows with purely discrete spectrum. Examples of positive entropy are given by Bernoulli flows. M.Ratner has shown that the classical horocycle flow H is LB, while $H \times H$ is not; this provides the first "natural" non-LB example ([R1],[R2]).

§3. Loosely Bernoulli transformations. While it will not be possible here to give a detailed account of these developments, we would like to give some of the definitions, at least in the finite entropy case, and briefly to discuss the techniques. Let $\mathbb{I}$ be a finite set, and α,β two functions from $\{1,\dots,N\}$ to $\mathbb{I}$. By $d_N(\alpha,\beta)$ is meant $N^{-1}|\{n : \alpha(n) \neq \beta(n)\}|$, where $|\dots|$ here means cardinality. This is often called the Hamming distance. $f_N(\alpha,\beta)$ is defined by setting $1 - f_N(\alpha,\beta)$ equal to $1/N$ times the maximal integer n for which there are positive integers $j_1 < \dots < j_n$ and $k_1 < \dots < k_n$ with $\alpha(j_i) = \beta(k_i)$, $i = 1,\dots,n$. Thus $f_N \le d_N$. For example, if $\alpha = (a,b,a,b,a,b)$ and $\beta = b,a,b,a,b,a)$, then $d_6(\alpha,\beta) = 1$ but $f_6(\alpha,\beta) = 1/6$.

Now let T be a finite entropy transformation on (X,μ), and $P = \{P_i : i \in \mathbb{I}\}$ a finite partition. Define $d_N^P(x,y) = d_N(P_1^N(x), P_1^N(y))$, where $P_1^N(x) = (P(x), P(Tx),\dots, P(T^N x))$, and $P(x)$ is that i for which $x \in P_i$. Similarly, $f_N^P(x,y) = f_N(P_1^N(x), P_1^N(y))$. Recall from [0] that (T,P) is said to be very weakly Bernoulli (VWB) if, given $\varepsilon > 0$, $\exists N_o$ so that if $N > N_o$ then there is a set G measurable with respect to $\bigvee_{j=-\infty}^{0} T^{-j}P$ and $\mu(G) > 1 - \varepsilon$, so that if $A \subset G$ is measurable with respect to $\bigvee_{i=-\infty}^{0} T^{-j}P$, and $\mu(A) > 0$, then there is a measure ν on $X \times X$ with right and left marginals μ_A and μ, respectively, and such that $\int_{X\times X} d_N^P(x,y)d\nu < \varepsilon$. Similarly, we say (T,P) is loosely Bernoulli (LB) if the same condition holds, but with f_N^P instead of d_N^P. We say T is LB if (T,P) is LB for every P.

For entropy zero the definition may be simplified: (T,P) is LB and of entropy 0 if and only if given $\varepsilon > 0$ $\exists N_o$ s.t. if $N > N_o$ then $\exists A$ with $\mu(A) > 1 - \varepsilon$ and $f_n^P(x,y) < \varepsilon$ for all $x,y \in A$. (By contrast, if (T,P) is VWB and of entropy zero, then P must be the trivial partition).

The definition of LB, like that of VWB, is mainly useful for checking examples and counterexamples. In order to prove that any two LB transformations of the same entropy class are isomorphic, one introduces another characterization, "finitely fixed," analogous to Ornstein's "finitely determined." Let (T,P) be a transformation and partition on (X,μ), and (S,Q) on (Y,ν), with both P and Q indexed by $\mathbb{I}$. By $\bar{d}_N((T,P),(S,Q))$ we mean $\inf\{\int d_N(P_1^N(x), Q_1^N(y))d\rho(x,y) : \rho$ any measure on $X \times Y$ with marginals μ and $\nu\}$. $\bar{d}_N$ increases with N, and $\bar{d}$ is defined as its limit. $\bar{f}_N$ and $\bar{f}$ are defined similarly.

Definition. (T,P) is said to be finitely determined (FD) if, given $\varepsilon > 0$, there exist N and δ such that if $h(T,P) < h(S,Q) + \delta$ and $\bar{d}_N((T,P),(S,Q)) < \delta$ then $\bar{d}((T,P),(S,Q)) < \varepsilon$. Here, as before, h means entropy. The use of $\bar{d}_N$ here is merely one of many possible ways of describing "closeness in finite distribution," but this way will be convenient for generalization later. On the other hand, the use of $\bar{d}$ is essential. We define finitely fixed (FF) in an analogous way, substituting $\bar{f}$ for $\bar{d}$ (but note: $\bar{d}_N$ is left unchanged!).

The proof of an equivalence-analogue of Sinai's Theorem and of Ornstein's Theorem may now be carried out for transformations T having a generating partition P such that (T,P) is FF. Analogously to the FD case, it turns out that then (T,R) is FF for all partitions R; and that (T,P) is FF if and only if (T,P) is FD.

§4. Onward to $n \geq 2$. The first problem one encounters in attempting to generalize to n-flows with $n \geq 2$, is that we have not yet asked the right question. In fact, D.Rudolph has shown in [Ru 2] that any two n-flows, $n \geq 2$, are equivalent! In order to get a nontrivial theory, one must restrict the sorts of reparametrizing maps which are used. It turns out that there is a natural class to use. The reason this does not arise for $n = 1$ is that all reparametrization maps for flows are in this natural class. We will not describe the class precisely, since the description is messy, and we will not need it in full generality. However, a basic result due to Rudolph [Ru 3] says that if ϕ is any n-flow and τ one of these natural reparametrizing maps for ϕ, then there exists another reparametrizing map σ such that $\sigma(x,\cdot)$ is a C^∞ diffeomorphism with $\sigma'(\cdot,0)$ and $\sigma'^{-1}(\cdot,0)$ uniformly bounded, and such that ϕ_τ is isomorphic to ϕ_σ. Thus, we may restrict our considerations to reparametrization maps satisfying these smoothness and boundedness conditions; we call these tempered, and will use terms like "tempered reparametrization", "tempered equivalence".

An analog of Abramov's formula has been proven for tempered reparametrization maps of n-flows, by D.Nadler [N] :

$$h(\phi) = (\int |\det \tau'(\cdot,0)| d\mu) h(\phi_\tau) .$$

Therefore there are at least three classes of flows which are in different tempered equivalence classes: those of zero entropy, positive but finite entropy, and infinite entropy. One is irresistably tempted to try to generalize the LB theory and examples to tempered equivalence of n-flows.

One strategy would be to try to follow the pattern of the 1-dimensional proof. Thus, one would need to make some generalization of the Ambrose theorem to reduce to a discrete situation. Katok, in [Ka3], has proven a theorem which certainly seems to be the right generalization: one can introduce a notion of a "special n-flow over a $\mathbb{Z}^n$ action," which is a generalization of a flow built over a transformation and under a constant function. The theorem then may be phrased: for every n-flow ϕ there is a tempered τ such that ϕ_τ is one of these special n-flows. However, the question when two $\mathbb{Z}^n$ actions can give rise this manner to isomorphic n-flows, we consider unmanageable: nothing as elegant as Kakutani equivalence appears to be available.

Another strategy would be to try to change the 1-dimensional definitions, proofs, and constructions, so that they deal directly with flows, rather than allowing the Ambrose theorem to intervene. Then one could try to generalize the proofs to n-flows. This approach has proven successful, and we shall describe it briefly. As before, we stick to the finite entropy case.

<u>§5. Tempered equivalence of n-flows of zero entropy.</u> Here is how the $\bar{d}$ and $\bar{f}$ metrics are carried over to flows, and even to n-flows. The work described in this section is carried out in [F-N].

Let D be a closed cell in $\mathbb{R}^n$ and let $|D|$ denote the Lebesgue measure of D . Let $\mathbb{I}$ be a finite set. Let $\alpha,\beta : D \to \mathbb{I}$ be measurable.

<u>Definition</u>: $d_D(\alpha,\beta) = \frac{1}{|D|} |\{v : \alpha(v) \neq \beta(v)\}|$.

Let $\mathcal{D}_D$ be the set of C^∞ self-diffeomorphism of D which are the identity in a neighborhood of the boundary of D . Let $\|A\|$ denote the operator norm of the $n \times n$ matrix A when $\mathbb{R}^n$ is given the sup norm. For a matrix-valued function λ with domain D, let $\|\lambda\|_\infty = \sup_{v\in D} \|\lambda(v)\|$.

<u>Definition</u>. $f_D(\alpha,\beta) = \inf_{h\in\mathcal{D}_D} [d_D(\alpha h, \beta) + \|h' - I\|_\infty]$.

<u>Definition</u>. If ϕ is an n-flow and $\mathcal{P} = \{P_i : i \in \mathbb{I}\}$ a finite partition, then we define a function $\mathcal{P}_D(x) : D \to \mathbb{I}$. $\mathcal{P}_D(x)(v)$ is that i for which $\phi_v x \in P_i$. Now define $d_D^{\mathcal{P}}(x,y) = d_D(\mathcal{P}_D(x), \mathcal{P}_D(y))$. Similarly, $f_D^{\mathcal{P}}(x,y) = F_D(\mathcal{P}_D(x), \mathcal{P}_D(y))$.

There is now an obvious definition of VWB for flows available. Unfortunately, it does not work; <u>any</u> flow ϕ has a generating partition $\mathcal{P}$ for which $(\phi,\mathcal{P})$ is <u>not</u> VWB in this sense. The obvious definition of LB for flows has a similar weakness. However, if we stick to flows of zero entropy, then the simpler version of the LB definition for zero entropy transformations can be transferred to flows, and even to n-flows. Here it is.

<u>Definition</u>. $(\phi,\mathcal{P})$ will be called <u>loosely Kronecker</u> (LK) if given $\varepsilon > 0$ $\exists$ N_0 such that if $N > N_0$ then there is a set H with $\mu(H) > 1 - \varepsilon$ and, if $x,y \in H$, $f_{C(N)}^{\mathcal{P}}(x,y) < \varepsilon$. Here $C(N)$ is the cube whose corners all have coordinates $\pm N$. The n-flow ϕ is called LK if $(\phi,\mathcal{P})$ is LK for all partitions $\mathcal{P}$. Any LK n-flow has zero entropy. The property LK is invariant under tempered reparametrization. Any n-flow with purely discrete spectrum is LK. Consequently, for $n = 1$, any LB flow of zero entropy is LK. Once we know that any two LK flows are linked by a tempered equivalence, it will follow that any LK flow is LB. One may make non LK n-flows of zero entropy by imitating the construction in [F1].

The most difficult task in this circle of ideas is to show that any two LK flows are linked by a tempered equivalence. One must introduce a flow analogue of FF. For the zero entropy case this is easy: say $(\phi,\mathcal{P})$ is FF if, given $\varepsilon > 0$, $\exists$ δ and N such that if $h(\psi,\mathcal{Q}) > h(\phi,\mathcal{P}) - \delta$ and $\bar{d}_{C(N)}((\psi,\mathcal{Q}),(\phi,\mathcal{P})) < \delta$ then $\bar{f}((\psi,\mathcal{Q}),(\phi,\mathcal{P})) < \varepsilon$. We have not defined $\bar{d}_{C(N)}$ and $\bar{f}$, but it should by now be obvious how to define them, in analogy with the discrete case. Then it may be shown that any two FF n-flows of zero entropy are linked by a tempered equivalence. The proof follows the model of Ornstein's Isomorphism Theorem, proving on the way the appropriate Sinai theorem. While in some ways it is technically quite difficult, the absence of entropy considerations makes important simplifications.

Finally, of course, it is shown that $(\phi,\mathcal{P})$ is FF of entropy zero if and only if $(\phi,\mathcal{P})$ is LK.

§6. r-entropy and the positive entropy case. This leads to interesting new problems and solutions. The first problem is to see how even to prove the isomorphism theorem for Bernoulli flows in a direct way, without appealing to the case of transformations. Even, what is the proper notion of FD ? Of course, there is an obvious definition, based on our continuous version of the $\bar{d}$ metric. Unfortunately, it fails even more badly than the continuous version of VWB : (ϕ,P) can never be FD. The trouble is that entropy for flows is not robust enough under small $\bar{d}$ changes. However, there is an approximate version of entropy, developed in [F2], which provides an effective substitute, which we shall describe.

The semimetric $d_D^P(x,y)$ enables us to speak of the P_D-diameter of a subset of X. Consider now a family $\mathcal{B}$ of disjoint, measurable sets in X , each having $P_{C(N)}$-diameter $\le r$, where r is fixed, $0 < r \le 1$. Suppose $\bigcup_{B\in\mathcal{B}} B$ has measure $> 1-\varepsilon$. We ask how small the number $h(\mathcal{B}) = -\sum_{B\in\mathcal{B}} \mu(B)\log\mu(B)$ can be. The r-entropy of (ϕ,P), written $h_r(\phi,P)$, is the infimum of numbers a such that for any $\varepsilon > 0$ there is some N_0 such that if $N > N_0$ then there is some family $\mathcal{B}$ as above with $\frac{1}{|C(N)|}h(\mathcal{B}) < a$. If instead of $h(\mathcal{B})$ we had used $\log|\mathcal{B}|$, the number $h_r(\phi,P)$ would have turned out to be exactly the same. For each P , $h_r(\phi,P)$ is a convex decreasing function of r whose value at 1 is zero and whose limit at 0 is the usual entropy $h(\phi,P)$. A "Macmillan" theorem holds for r-entropy: for any fixed r , if ε is sufficiently small, N sufficiently large, and $\frac{1}{|C(N)|}h(\mathcal{B})$ sufficiently close to $h_r(\phi,P)$, then, except for a set of small total measure, all the $\mu(B)$ have $\frac{-\log\mu(B)}{|C(B)|}$ close to $h_r(\phi,P)$.

Now we are ready to make the proper version of "finitely determined" for n-flows.

Definition. (ϕ,P) is said to be semifinitely determined (SFD) if, given $\varepsilon > 0$, there exist N, γ, and $\delta(r)$, $0 < r \le 1$, so that given any (ψ,Q) with Q indexed like P , and such that there is some r for which

(a) $h_r(\phi,P) < h_r(\psi,Q) + \gamma$,

(b) $\bar{d}_N((\phi,P),(\psi,Q)) < \delta(r)$,

then $\bar{d}((\phi,P),(\psi,Q)) < \varepsilon$.

It is shown in [F2] that if ϕ is a Bernoulli n-flow, then (ϕ,P) is always SFD, and that if ϕ and ψ are n-flows of equal entropy and both having SFD generators, then they are isomorphic. The consequent result, that any two Bernoulli n-flows of equal entropy are isomorphic, was already known; see Lind [L]. But our technique will be needed to provide a model for a tempered equivalence theorem.

It is clear, in general, how to proceed. We define a semifinitely fixed (SFF) n-flow by substituting $\bar{f}$ for $\bar{d}$ in the definition of SFD (but again: leaving $\bar{d}_N$ alone). The definition coincides with the definition of FF for zero entropy n-flows made in the last section. Then the theory carries over. First, if the n-flow ϕ has a generating partition $\mathcal{P}$ such that $(\phi,\mathcal{P})$ is SFF, then $(\phi,\mathcal{Q})$ is SFF for any $\mathcal{Q}$. Call such a ϕ SFF. The property is invariant under tempered equivalence, and any two SFF flows of positive and finite entropy are linked by a tempered equivalence. The Bernoulli n-flows provide SFF examples; and products of these with the non-SFF zero entropy n-flows of the last section provide non-SFF examples of positive finite entropy. This last work is still in the process of being written, by D.Ornstein and the present author.

Bibliography

[A] Abramov,P.:"Entropy of induced transformations," Dok.Akad. S.S.S.R., 128, No.4 (1959), 647-650 (in Russian).

[Am] Ambrose,W.: "Representation of ergodic flows," Ann.Math. 42 (1941), 723-739.

[F1] Feldman,J.: "New K-automorphism and a problem of Kakutani," Israel J. of Math., 24, No.1 (1976), 16-37.

[F2] Feldman,J.: "r-Entropy, equipartition, and Ornstein's isomorphism theorem in $\mathbb{R}^n$," preprint, U.C. Berkeley, 1978.

[F-N] Feldman,J. and Nadler,D.: "Reparametrization of n-flows of zero entropy," preprint, U.C. Berkeley, 1978.

[K] Kakutani,S.: "Induced measure-preserving transformations," Proc.Imp.Acad. Tokyo 19(1943), 635-641.

[Ka1] Katok,A.: "Time change, monotone equivalence, and standard dynamical systems," Dok.Akad.Nauk. S.S.S.R. 273 (1975), 789-792 (in Russian).

[Ka2] Katok,A.: "Monotone equivalence in ergodic theory," Izvestia Mat.Nauk.,v.41 (1977), 104-157 (in Russian).

[Ka3] Katok,A.: "The spectral representation theorem for multidimensional group actions," Asterisk 49 (1977) (Proceedings of Warsaw conference on Ergodic Theory).

[L] Lind,D.: "Locally compact measure-preserving flows," Advances in Math. 15 (1975), 175-193.

[N] Nadler,D.: "Abramov's formula for reparametrization of n-flows," preprint, U.C.Berkeley, 1978.

[O] Ornstein,D.: "Randomness, Ergodic Theory, and Dynamical Systems," Yale Mathematical Monographs No.5.

[R1] Ratner,M.: "Horocycle flows are loosely Bernoulli," to appear, Israel J.Math.

[R2] Ratner,M.: "The Cartesian square of the horocycle flow is not loosely Bernoulli," preprint, U.C.Berkeley, 1978.

[Ru1] Rudolph,D.: "Non-equivalence of measure-preserving transformations," lecture notes, Hebrew University, 1976.

[Ru2] Rudolph,D.: "A Dye Theorem for n-flows, n > 1," preprint, U.C. Berkeley, 1977.

[Ru3] Rudolph,D.: "An integrably Lipschitz reparametrization of an n-flow is isomorphic to some tempered reparametrization of the same n-flow," preprint, U.C.Berkeley, 1978.

[Sa] Satayev,E.: "An invariant of monotone equivalence which determines families of automorphisms which are monotone equivalent to a Bernoulli automorphism," Proc. Fourth Symp. on Information Theory, Part I, Moscow-Leningrad, 1976 (in Russian).

[W] Weiss,B.: "Equivalence of measure preserving transformations," lecture notes, Hebrew University, 1976.

Jacob Feldman
Dept. of Mathematics
970 Evans Hall
University of California
Berkeley, Calif. 94720
U.S.A.

FUNDAMENTAL HOMOMORPHISM OF NORMALIZER

GROUP OF ERGODIC TRANSFORMATION

Toshihiro HAMACHI and Motosige OSIKAWA

1. INTRODUCTION

Let (Ω, F, P) be a finite Lebesgue measure space and G be a group of countable automorphisms of Ω, where an automorphism means a measurable and invertible transformation ϕ satisfying $P(A) = 0$ if and only if $P(\phi A) = 0$ for $A \in F$. We say G is ergodic if every G-invariant set is a null set or a co-null set. An automorphism R of Ω is called a normalizer of G if $R\mathrm{Orb}_G(\omega) = \mathrm{Orb}_G(R\omega)$ a.e.ω, where $\mathrm{Orb}_G(\omega)$ is the orbit $\{ g\omega ; g \in G \}$. We denote by $N[G]$ the set of these R. The set of automorphisms ϕ such that $\phi\omega \in \mathrm{Orb}_G(\omega)$ a.e.ω is a subgroup of $N[G]$, which we call the full group of G and denote by $[G]$. It is known in [1] that an ergodic group G of countable automorphisms admits a σ-finite invariant measure μ equivalent to P, then every $R \in N[G]$ has a module $\frac{d\mu R}{d\mu}(\omega)$, which is a constant a.e.ω, and that the module is an invariant for the outer conjugacy of normalizers. On the other hand in von Neumann Algebra theory A.Connes and M. Takesaki [2] introduced a module of automorphisms of a type III factor. Here we give a measure theoretical version of Connes and Takesaki's module to $N[G]$ in the case that G does not necessarily admit an invariant measure as follows.

Let for $R \in N[G]$ $\widetilde{R}$ be the skew product automorphism defined by $\widetilde{R}(\omega, u) = (R\omega, u - \log\frac{dPR}{dP}(\omega))$, $(\omega, u) \in \Omega\times R$. Let $\zeta(\widetilde{G})$ be the measurable partition of $\Omega\times R$ which generates the σ-algebra consisting of all $\widetilde{G}$-invariant sets, where $\widetilde{G} = \{\widetilde{g} ; g \in G\}$. Since for $R \in N[G]$ $\widetilde{R}$ preserves $\zeta(\widetilde{G})$, $\widetilde{R}\zeta(\widetilde{G}) = \zeta(\widetilde{G})$, we can define the factor automorphism of $\widetilde{R}$ acting on the quotient measure space $\Omega\times R/\zeta(\widetilde{G})$. We call this factor automorphism a module of R and denote it by mod R. Module keeps the following properties :

(1) mod R is the identity map if $R \in [G]$.

(2) $\mathrm{mod}(R_1R_2) = \mathrm{mod}\, R_1 \cdot \mathrm{mod}\, R_2$ and $\mathrm{mod}(R^{-1}) = (\mathrm{mod}\, R)^{-1}$.

(3) If $R_1 \in N[G]$ and $R_2 \in N[G]$ are outer conjugate then mod R_1 is isomorphic to mod R_2.

(4) mod R commutes with the associated flow $\{\widetilde{T}_s\}_{s\in R}$ of G, which is the factor flow of $\{T_s\}_{s\in R}$, $T_s(\omega, u) = (\omega, u + s)$ for $(\omega, u) \in \Omega\times R$, acting on the quotient measure space $\Omega\times R/\zeta(\widetilde{G})$.

Since it is known that G admits an equivalent σ-finite invariant measure if and only if the associated flow $\{\widetilde{T}_s\}_{s\in R}$ is isomorphic to the translation, $R \ni u \longrightarrow u + s$ for $s \in R$, from (4) we have

(5) If G admits a σ-finite invariant measure μ equivalent to P then mod R of a normalizer R of G is isomorphic to the automorphism, $R \ni u \longrightarrow u + c$, where $c = \log\frac{d\mu R}{d\mu}(\omega)$ a.e.ω. If $\mu(\Omega) < \infty$ then mod R is the identity map.

(6) If G is of type III_λ, $0 < \lambda < 1$, that is, the associated flow $\{\tilde{T}_s\}_{s R}$ of G is isomorphic to the periodic flow, $[0,-\log\lambda) \ni u \longrightarrow u + s \pmod{-\log\lambda}$, then mod R of a normalizer R of G is isomorphic to an automorphism, $[0,-\log\lambda) \ni u \longrightarrow u + c \pmod{-\log\lambda}$ for some c.

(7) If G is of type III_1, that is, $\tilde{G}$ is ergodic, then mod R of a normalizer R of G is trivial.

Module is a mapping from N[G] to the set of flows commuting with the associated flow of G and is called the fundamental homomorphism. In [3] we introduced a complete metrizable topology on N[G]. In section 3 we shall prove that if G is generated by a single ergodic automorphism then mod R is the identity map if and only if R is in the closure of [G]. In section 4 we give examples of G, the module of whose Sakai's flip is the identity map. A.Connes and J.Woods [5] gave an approximately finite dimensional factor, the module of whose Sakai's flip is not the identity map.

2. EXAMPLES OF NORMALIZERS

Let (X , m) be a bilateral infinite direct product measure space $\left(\prod_{k=-\infty}^{\infty} \{0,1,2\}, \prod_{k=-\infty}^{\infty} m_k \right)$, where $m_k(0) = 1/1+\lambda+\eta$, $m_k(1) = \lambda/1+\lambda+\eta$ and $m_k(2) = \eta/1+\lambda+\eta$, $k=0,\pm 1,\pm 2,\ldots$, $0 < \lambda,\eta < 1$. Let Γ be the infinite direct product group of automorphisms of X, that is, the group generated by automorphisms g_n, $(g_n x)_n = x_n + 1 \pmod 2$ and $(g_n x)_k = x_k$ for $k \neq n$, for $x = (x_k) \in X$ and $n=0,\pm 1,\pm 2,\ldots$.

Let H be the group of automorphisms of X permutating finite coordinates , that is , the group generated by automorphisms h_{mn} , $(h_{mn}x)_m = x_n$, $(h_{mn}x)_n = x_m$ and $(h_{mn}x)_k = x_k$ for $k \neq m,n$, for $x = (x_k) \in X$ and $m,n=0,\pm 1,\pm 2,\ldots$. Let θ be the shift on X , $(\theta x)_k = x_{k+1}$, $k=0,\pm 1,\pm 2,\ldots$. Let for a positive integer N θ_N be an automorphism defined by $(\theta_N x)_k = x_{k+1}$ for $k=jN,jN+1,\ldots,jN+N-2$ and $(\theta_N x)_{jN+N-1} = x_{jN}$, $j=0,\pm 1,\pm 2,\ldots$. We racall the definition of outer periodicity of normalizers. If a normalizer R of G satisfies $R^p \in [G]$ for some positive integer p then we say R is outer periodic, and if not , R is outer aperiodic.

<u>Example (II_1).</u>

H is an ergodic group of finite measure preserving automorphisms. θ_N (resp. θ) is outer periodic (resp. aperiodic) normalizers of H .

<u>Example (II_∞).</u>

We assume that $\log\lambda$ and $\log\eta$ are rationally independent. Then the group $\tilde{\Gamma}$ consisting of skew product automorphisms $\tilde{g}$, $\tilde{g}(x , u) = (gx , u - \log\frac{dmg}{dm}(x))$, $(x , u) \in X\times R$, for $g \in \Gamma$, is ergodic and preserves a σ-finite infinite measure $dm(x)\times e^u du$. Automorphisms $\tilde{\theta}_N$ (resp. $\tilde{\theta}$) defined by $\tilde{\theta}_N(x , u) = (\theta_N x , u)$ (resp. $\tilde{\theta}(x , u) = (\theta x , u)$) is outer periodic (resp. outer aperiodic) normalizers of $\tilde{\Gamma}$ and their modules are the identity map. An automorphism R defined by $R(x , u) = (x , u + c)$, where c is a constant , is a normalizer of $\widetilde{\Gamma}$ and mod R is isomorphic to an automorphism $R \ni u \longrightarrow u + c$.

Example (III_λ) ($0< \lambda <1$).

We assume that $\log\lambda$ and $\log\eta$ are rationally independent. Then the group generated by the group $\widetilde{\Gamma}$ defined in the case (II_∞) and an automorphism $X\times R \ni (x , u) \longrightarrow (x , u + \log\lambda)$, is of type III_λ . $\widetilde{\theta}_N$ (resp. $\widetilde{\theta}$) defined in the case (II_∞) is an outer periodic (resp. aperiodic) normalizer of this group, whose module is the identity map. R defined in the case (II_∞) is outer periodic if c and $\log\lambda$ are rationally independent and outer aperiodic if not. The module mod R is isomorphic to an automorphism $[0,-\log\lambda) \ni u \longrightarrow u + c$ (mod $(-\log\lambda)$).

Example (III_1).

We assume that $\log\lambda$ and $\log\eta$ are rationally independent Then the group $\widetilde{\Gamma}$ defined in the case (II_∞) and automorphisms $X\times R \ni (x , u) \longrightarrow (x , u + \log\lambda)$ and $(x , u) \longrightarrow (x , u + \log\eta)$ is of type III_1 . $\widetilde{\theta}_N$ (resp. $\widetilde{\theta}$) defined in the case (II_∞) is an outer periodic (resp. aperiodic) normalizer of this group.

Example (III_0).

We assume that $\log\lambda$ and $\log\eta$ are rationally independent.

Let Q be an ergodic automorphism of a Lebesgue measure space (Y, ν) and $f(y)$ be a measurable function such that $f(y) > \delta > 0$. The group generated by automorphisms $X\times R\times Y \ni (x, u, y) \longrightarrow (gx, u - \log\frac{dmg}{dm}(x), y)$ for $g \in \Gamma$ and an automorphism $X\times R\times Y \ni (x, u, y) \longrightarrow (x, u - \log\frac{d\nu Q}{d\nu}(y) - f(y), Qy)$ is type III_0, in fact its associated flow is isomorphic to the flow built under the function $f(y)$ with base automorphism Q. An automorphism $X\times R\times Y \ni (x, u, y) \longrightarrow (\theta_N x, u, y)$ (resp. $(x, u, y) \longrightarrow (\theta x, u, y)$) is an outer periodic (resp. aperiodic) normalizer of this group, whose module is the identity map. An automorphism $X\times R\times Y \ni (x, u, y) \longrightarrow (x, u - c, y)$, where c is a constant, is an outer aperiodic normalizer ($c \neq 0$) and its module is isomorphic to an automorphism $\widetilde{T}_c$, where $\{\widetilde{T}_s\}_{s\in R}$ is the associated flow of the group. In particular case that $f(y) = \text{constant} = K$ every automorphism commuting with a flow built under a constant function K with base automorphism Q is of the form ; $\psi(y, u) = (Vy, u + c)$ for (y, u) in $Y\times[0,K)$, where V is an automorphism commuting with Q and c a constant. Set $R(x, u, y) = (x, u - c, Vy)$ then $\mathrm{mod}\, R$ is isomorphic to ψ.

3. CLASS OF NORMALIZERS WHOSE MODULES ARE THE IDENTITY MAP.

In [3] we introduced a topology on $N[G]$ in order to give a proof of a Krieger's cohomology lemma [4]. It is as follows. Let for $R \in N[G]$ $R_P\xi(\omega) = \xi(R^{-1}\omega)\frac{dPR^{-1}}{dP}(\omega)$ for $\xi \in L^1(P)$. Then R_P is a positive isometry on $L^1(P)$. For $R_n \in N[G]$, $n=1,2,\ldots$ and $R \in N[G]$, we say R_n converges to R if and only if

(1) $(R_n)_P\xi \longrightarrow R_P\xi$ as $n \longrightarrow \infty$ in $L^1(P)$-norm for $\xi \in L^1(P)$

and

(2) $P(\ \omega \in \Omega\ ;\ R_n g R_n^{-1}\omega \neq RgR^{-1}\omega\ ) \longrightarrow 0$ as $n \to \infty$ for all $g \in [G]$.

Then we have the following properties.

(3) The topology does not depend on the choice of a finite measure equivalent to P .

(4) $N[G]$ is a topological group with respect to the topology.

(5) The topology is compatible with the following complete metric d ; for $R \in N[G]$ and $S \in N[G]$,

$$d(R,S) = \sum_{k=1}^{\infty} \frac{1}{2^k} \times \frac{\|(R_P - S_P)\xi_k\|_{L^1(P)} + \|(R_P^{-1} - S_P^{-1})\xi_k\|_{L^1(P)}}{1 + \|(R_P - S_P)\xi_k\|_{L^1(P)} + \|(R_P^{-1} - S_P^{-1})\xi_k\|_{L^1(P)}}$$

$$+ \sum_{k=1}^{\infty} \frac{1}{2^k} \times \frac{P(\omega\ \Omega; Rg_kR^{-1}\omega \neq Sg_kS^{-1}\omega) + P(\omega\ \Omega; R^{-1}g_kR\omega \neq S^{-1}g_kS\omega)}{1 + P(\omega\ \Omega; Rg_kR^{-1}\omega \neq Sg_kS^{-1}\omega) + P(\omega\ \Omega; R^{-1}g_kR\omega \neq S^{-1}g_kS\omega)}$$

where $\{\ \xi_k\ ,\ k=1,2,\ldots\}$ is a countable dense set in $L^1(P)$ and $G = \{\ g_k\ ,\ k=1,2,\ldots\ \}$.

(6) (1) is equivalent to the conditions :

$P(R_nA \triangle RA) \longrightarrow 0$ as $n \longrightarrow \infty$ for $A \in F$ and

$\frac{dPR_n^{-1}}{dP}(\omega) \longrightarrow \frac{dPR^{-1}}{dP}(\omega)$ as $n \longrightarrow \infty$ in the $L^1(P)$-norm.

If for a group G of countable automorphisms there exists an automorphism T such that $[G] = [\ \{T^n\}_{n\in Z}]$ then we say G is approximately finite . We write $[T]$ (resp. $N[T]$) instead of $[\ \{T^n\}_{n\in Z}]$ (resp. $N[\{T^n\}_{n\in Z}]$).

<u>Theorem.</u>

Let T be an ergodic automorphism of a finite Lebesgue

measure space (Ω, F, P) and let $R \in N[T]$. Then mod R is the identity map if and only if $R \in \overline{[T]}$, where $\overline{[T]}$ is the closure of $[T]$ with respect to the topology above mentioned.

<u>Proof.</u> Suppose $R_n \in [T]$ converges to $R \in N[T]$ as $n \longrightarrow \infty$. For a proof of the necessity of the theorem it is enough to show that if a bounded measurable function $f(\omega,u)$ is $\widetilde{T}$-invariant then it is also $\widetilde{R}$-invariant. For this we show that for a bounded measurable function $\phi(\omega)$, $|\phi(\omega)| < K$, and an interval $I = [a,b]$,

$$J = \int\int \phi(\omega)\chi_I(u)f(R_n\omega, u-\log\frac{dPR_n}{dP}(\omega))e^u du dP(\omega)$$
$$- \int\int \phi(\omega)\chi_I(u)f(R\omega, u-\log\frac{dPR}{dP}(\omega))e^u du dP(\omega)$$

converges to 0 as $n \longrightarrow \infty$. Let $|f(\omega,u)| < L$. We have

$$J = \int\int \phi(R_n^{-1}\omega)\chi_I(u-\log\frac{dPR_n^{-1}}{dP}(\omega))f(\omega,u)e^u du dP(\omega)$$
$$- \int\int \phi(R^{-1}\omega)\chi_I(u-\log\frac{dPR^{-1}}{dP}(\omega))f(\omega,u)e^u du dP(\omega)$$
$$= \int\int \phi(R_n^{-1}\omega)f(\omega,u)\{\chi_I(u-\log\frac{dPR_n^{-1}}{dP}(\omega))-\chi_I(u-\log\frac{dPR^{-1}}{dP}(\omega)\}e^u du$$
$$\times dP(\omega) - \int\int \{\phi(R_n^{-1}\omega)-\phi(R^{-1}\omega)\}\chi_I(u-\log\frac{dPR^{-1}}{dP}(\omega))f(\omega,u)e^u du dP(\omega)$$
$$= J_1 + J_2 .$$

And we have

$$|J_1| \leqq KL \int\int|\chi_I(u-\log\frac{dPR_n^{-1}}{dP_1}(\omega))-\chi_I(u-\log\frac{dPR^{-1}}{dP}(\omega)|e^u du dP(\omega)$$
$$\leqq KL\int\{|\exp(a+\log\frac{dPR_n^{-1}}{dP}(\omega))-\exp(a+\log\frac{dPR^{-1}}{dP}(\omega))|$$
$$+ |\exp(b+\log\frac{dPR_n^{-1}}{dP}(\omega))-\exp(b+\log\frac{dPR^{-1}}{dP}(\omega))|\}dP(\omega)$$
$$= KL(e^a+e^b)\int|\frac{dPR_n^{-1}}{dP}(\omega) - \frac{dPR^{-1}}{dP}(\omega)|dP(\omega)$$
$$\longrightarrow 0 \text{ as } n \longrightarrow \infty .$$

Set $g(\omega) = \int\chi_I(u-\log\frac{dPR^{-1}}{dP}(\omega))f(\omega,u)e^u du$ then $g(\omega) \in L^1(P)$

and we have

$$|J_2| = |\int \phi(R_n^{-1}\omega)g(\omega)dP(\omega) - \int \phi(R^{-1}\omega)g(\omega)dP(\omega)|$$

$$= |\int \phi(\omega)g(R_n\omega)dP(R_n\omega) - \int \phi(\omega)g(R\omega)dP(R\omega)|$$

$$= |\int \phi(\omega)\{g(R_n\omega)\frac{dPR_n}{dP}(\omega)-g(R\omega)\frac{dPR}{dP}(\omega)\}dP(\omega)|$$

$$\leqq K \|(R_n^{-1})_P g - (R^{-1})_P g\|_{L^1(P)}$$

$$\longrightarrow 0 \quad \text{as} \quad n \longrightarrow \infty .$$

Since $f(\omega,u)$ is $\widetilde{T}$-invariant, we have a desired conclusion :

$$\int\int\phi(\omega)\chi_I(u)f(\omega,u)e^u du dP(\omega) = \int\int \phi(\omega)\chi_I(u)f(R\omega,u-\log\frac{dPR}{dP}(\omega)) \times e^u du dP(\omega) ,$$

that is , $f(\omega,u) = f(R\omega,u-\log\frac{dPR}{dP}(\omega))$ for a.e.(ω,u) .

To prove the converse, we use the terminology "[T]-array". For subsets A and B a one to one map g from A onto B is called a [T]-map from A onto B if Pg is equivalent to P on A and $g\omega \in \mathrm{Orb}_T(\omega)$ for a.e.ω . A [T]-array of A , $P(A) > 0$, consists of a subsets B of A and [T]-maps g_i , $i=0,1,\ldots,n-1$ such that g_0 = the identity map and $\{ g_iB , i=0,1,\ldots,n-1 \}$ is a partition of A . We denote such a [T]-array ζ by $\zeta = \{ B ; g_i , i=0,1,\ldots,n-1 \}$. Assign to a permutation π of $\{0,1,\ldots,n-1\}$ the automorphism $g(\pi)$ by $g(\pi)\omega = g_{\pi(j)}g_j^{-1}\omega$ for $\omega \in g_jB$, $j=0,1,\ldots,n-1$ and denote by $G(\zeta)$ the finite group of these automorphisms $g(\pi)$ and by $G(\zeta)\omega$ the orbit $\{ g\omega ; g \in G(\zeta)\}$. Let $\zeta = \{ B ; g_i , i=0,1,\ldots,n-1 \}$ be a [T]-array of A and $\eta = \{ C ; f_j , j=0,1,\ldots,m-1 \}$ be a [T]-array of B , then we denote by $\zeta\times\eta$ the [T]-array $\{ C ; g_if_j , i=0,1,\ldots,n-1 , j=0,1,\ldots,m-1\}$ of A . We say that a [T]-array ζ_1 of a set is a refinement of a [T]-array ζ of the same set if $\zeta_1 = \zeta\times\eta$ for some [T]-array η.

Proof of the sufficiency of the Theorem.

Let $\zeta_n = \{ A_n ; g_i^{(n)} , i=0,1,\ldots,L_n-1 \}$, $n=1,2,\ldots$ be a sequence of [T]-arrays such that ζ_{n+1} is a refinement of ζ_n, $n=1,2,\ldots$, $T\omega \in \bigcup_{n=1}^{\infty} G(\zeta_n)\omega$, a.e.$\omega$ and $\{ g_i^{(n)}A_n , i=0,1,\ldots, L_n-1 , n=1,2,\ldots\}$ is a countable dense base of F. It is easy to see that for a sequence $\{\varepsilon_n\}$ with $\varepsilon_n \downarrow 0$ there exist automorphisms $R_n \in [T]$ such that $R_n g_i^{(n)} A_n = Rg_i^{(n)} A_n$ for $i=0,1,\ldots,L_n-1$ and $R_n g_i^{(n)} R_n^{-1}\omega = Rg_i^{(n)} R^{-1}\omega$, a.e.$\omega \in RA_n$ for $i=0,1,\ldots,L_n-1$ and $\exp(-\varepsilon_n) < \frac{dPR_n}{dP}(\omega)/\frac{dPR}{dP}(\omega) < \exp(\varepsilon_n)$, a.e.$\omega$, $n=1,2,\ldots$, then R_n converges to R as $n \to \infty$. By assuming that mod R is the identity map we prove the existence of such automorphisms R_n in each case.

Case (II_1). T is ergodic finite measure preserving with an invariant measure P. Since $P(RA_n) = P(A_n)$, there exists a [T]-map R_n from A_n onto RA_n. Set $R_n\omega = Rg_i^{(n)}R^{-1}R_n(g_i^{(n)})^{-1}\omega$ for $\omega \in g_i^{(n)}A_n$. Since $Rg_i^{(n)}R^{-1}\omega \in Orb_T(\omega)$, R_n is a [T]-map from $g_i^{(n)}A_n$ onto $Rg_i^{(n)}A_n$ such that $R_n g_i^{(n)} R_n^{-1}\omega = Rg_i^{(n)}R^{-1}\omega$, $i=0,1,\ldots,L_n-1$.

Case (III_λ). T is of type III_λ, $0 < \lambda < 1$, and P is a finite admissible measure, that is, the subgroup $\{ g \in [T] ; Pg = P \}$ is ergodic and $\frac{dPg}{dP}(\omega) \in \{ \lambda^n , n=0,\pm1,\pm2,\ldots\}$ a.e.ω for $g \in [T]$. By the assumption we have $\frac{dPR}{dP}(\omega) \in \{ \lambda^n , n=$

$0, \pm 1, \pm 2, \ldots\}$ a.e.ω . Then we have a countable partition $\{ A_{n,j} , j=1,2,\ldots \}$ of A_n such that $\frac{dPR}{dP}(\omega)$ = constant for $\omega \in g_i^{(n)} A_{n,j}$ and $\frac{dPg_i^{(n)}}{dP}(\omega)$ = constant for $\omega \in A_{n,j}$, $i=0,1,\ldots,L_n-1$, $j=1,2,\ldots$. These constants depend on n, i and j . Then there exists a [T]-map R_n from A_n onto RA_n such that $R_n A_{n,j} = RA_{n,j}$ and $\frac{dPR_n}{dP}(\omega) = \frac{dPR}{dP}(\omega)$ for a.e.$\omega \in A_n$. Since $Rg_i^{(n)}R^{-1}\omega \in \mathrm{Orb}_T(\omega)$ a.e.ω , we can find a [T]-map from $g_i^{(n)}A_n$ onto $Rg_i^{(n)}A_n$ such that $R_n g_i^{(n)} A_{n,j} = Rg_i^{(n)} A_{n,j}$, $R_n g_i^{(n)} R_n^{-1}\omega = Rg_i^{(n)}R^{-1}\omega$ for a.e.$\omega \in RA_n$ and $\frac{dPR_n}{dP}(\omega) = \frac{dPR}{dP}(\omega)$ a.e.ω .

<u>Case (III$_0$).</u> T is of type III$_0$. W.Krieger showed in [4] that there exist an ergodic automorphism Q of a Lebesgue measure space (Y , ν) and a measurable function $f(y) > \delta > 0$ on Y such that T is weakly equivalent to the group generated by automorphisms $\bar{g} : X\times R\times Y \ni (x,u,y) \longrightarrow (gx, u - \log\frac{dmg}{dm}(x), y)$, for g in the group Γ defined in section 2 , and an automorphism $Q_f : X\times R\times Y \ni (x,u,y) \longrightarrow (x, u - \log\frac{d\nu Q}{d\nu}(y) - f(y), Qy)$. Then we may assume that $\Omega = X\times R\times Y$, $dP = dm\times e^u du\times d\nu$ and [T] is the full group of the above group. We have a countable partition $\{ A_{n,j} , j=1,2,\ldots\}$ such that

$c_{i,j}^{(n)}\exp(-\frac{\varepsilon_n}{6}) < \frac{dPg_i^{(n)}}{dP}(\omega) < c_{i,j}^{(n)}\exp(\frac{\varepsilon_n}{6})$ for $\omega \in A_{n,j}$ and

$d_{i,j}^{(n)}\exp(-\frac{\varepsilon_n}{6}) < \frac{dPR}{dP}(\omega) < d_{i,j}^{(n)}\exp(\frac{\varepsilon_n}{6})$ for $\omega \in g_i^{(n)}A_{n,j}$, $i=0,1,\ldots,L_n-1$, $j=1,2,\ldots$, where $c_{i,j}^{(n)}$ and $d_{i,j}^{(n)}$ are positive constants. By the assumption there exists an automorphism g in the full group $[\{Q_f^n , n=0,\pm1,\pm2,\ldots\}]$ such that $\frac{dPR}{dP}(\omega) = \frac{dPg}{dP}(\omega)$ a.e.ω . Since

$dm\times e^u du((RA_{n,j})_y) = dm\times e^u du((gA_{n,j})_y)$ for a.e.y , where A_y means the y-section of a set A , there exists an automorphism h in the full group $[\{\bar{g}\ ,\ g \in \Gamma\}]$ such that $hRA_{n,j} = gA_{n,j}$. Therefore we have a [T]-map R_n such that $R_n A_{n,j} = RA_{n,j}$, $j=1,2,\ldots$ and $\frac{dPR_n}{dP}(\omega) = \frac{dPR}{dP}(\omega)$ for $\omega \in A_n$. Since $Rg_i^{(n)}R^{-1}\omega \in \mathrm{Orb}_T(\omega)$, we can find a [T]-map R_n from $g_i^{(n)}A_n$ onto $Rg_i^{(n)}A_n$ such that $R_n g_i^{(n)}A_{n,j} = Rg_i^{(n)}A_{n,j}$, $R_n g_i^{(n)}R_n^{-1}\omega = Rg_i^{(n)}R^{-1}\omega$ a.e.$\omega \in RA_n$,$i=0,1,\ldots,L_n-1$, $j=1,2,\ldots$ and $\exp(-\varepsilon_n) < \frac{dPR_n}{dP}(\omega)/\frac{dPR}{dP}(\omega) < \exp(\varepsilon_n)$ a.e.ω .

<u>Case (III_1).</u> T is of type III_1 . Let $\{ A_{n,j}\ ,\ j=1,2,\ldots \}$ be a countable partition of A_n such that

$c_{i,j}^{(n)}\exp(-\frac{\varepsilon_n}{7}) < \frac{dPg_i^{(n)}}{dP}(\omega) < c_{i,j}^{(n)}\exp(\frac{\varepsilon_n}{7})$ for $\omega \in A_{n,j}$ and $d_{i,j}^{(n)}\exp(-\frac{\varepsilon_n}{7}) < \frac{dPR}{dP}(\omega) < d_{i,j}^{(n)}\exp(\frac{\varepsilon_n}{7})$ for $\omega \in g_i^{(n)}A_{n,j}$, $i=0,1,\ldots,L_n-1$, $j=1,2,\ldots$, where $c_{i,j}^{(n)}$ and $d_{i,j}^{(n)}$ are positive constants. From the property of type III_1 there exists a [T]-map R_n from A_n onto RA_n such that $R_n A_{n,j} = RA_{n,j}$ and $\frac{P(RA_{n,j})}{P(A_{n,j})}\exp(-\frac{\varepsilon_n}{7}) < \frac{dPR_n}{dP}(\omega) < \frac{P(RA_{n,j})}{P(A_{n,j})}\exp(\frac{\varepsilon_n}{7})$ for a.e.$\omega \in A_{n,j}$, $j=1,2,\ldots$.

Since $Rg_i^{(n)}R^{-1}\omega \in \mathrm{Orb}_T(\omega)$, we can find a [T]-map R_n from $g_i^{(n)}A_n$ onto $Rg_i^{(n)}A_n$ such that $R_n g_i^{(n)}A_{n,j} = Rg_i^{(n)}A_{n,j}$, $R_n g_i^{(n)}R_n^{-1}\omega = Rg_i^{(n)}R^{-1}\omega$ for a.e.$\omega \in RA_n$ and $\exp(-\varepsilon_n) < \frac{dPR_n}{dP}(\omega)/\frac{dPR}{dP}(\omega) < \exp(\varepsilon_n)$ a.e.ω .

<u>Case (II_∞).</u> T is ergodic and σ-finite infinite measure preserving with an invariant measure P . Let $\{ \Omega_n\ ,\ n=1,2,\ldots\}$ be a countable partition of Ω with $0 < P(\Omega_n) < \infty$.

Since $P(R\Omega_n) = P(\Omega_n)$, n=1,2,... , there exists an automorphism $g \in [T]$ such that $gR\Omega_n = \Omega_n$, n=1,2,... . By the conclusion of case (II_1) , gR is in the closure $\overline{[T_{\Omega_n}]}$ on Ω_n , n=1,2,... , where T_{Ω_n} is the induced automorphism of T on Ω_n. Therefore $gR \in \overline{[T]}$ and $R \in g^{-1}\overline{[T]} = \overline{[T]}$.

4. CLASS OF ERGODIC AUTOMORPHISMS, THE MODULE OF WHOSE SAKAI'S FLIP IS THE IDENTITY MAP.

For an ergodic group of countable automorphisms of (Ω , $\mathcal{F}$, P) $G\times G$ is the group consisting of automorphisms $g\times g'$ of $\Omega\times\Omega$ defined by $g\times g'(\omega,\omega') = (g\omega,g'\omega')$ for g and g' in G . An automorphism σ of $\Omega\times\Omega$ defined by $\sigma(\omega,\omega') = (\omega',\omega)$ is a normalizer of $G\times G$ and called Sakai's flip. Let $\{\tilde{T}_s\}_{s\in R}$ be the associated flow of G acting on the quotient measure space $W = \Omega\times R/\zeta(\tilde{G})$. Let $\zeta(\{\tilde{T}_s\times\tilde{T}_{-s}\}_{s\in R})$ be a measurable partition of $W\times W$ which generates the σ-algebra consisting of all $\{\tilde{T}_s\times\tilde{T}_{-s}\}_{s\in R}$-invariant measurable sets. Set $\sigma_W(w,w') = (w',w)$ for $(w,w') \in W\times W$ and let $\bar{\sigma}_W$ be the factor automorphism of σ_W defined on the quotient measure space $W\times W/\zeta(\{\tilde{T}_s\times\tilde{T}_{-s}\}_{s\in R})$. Then it can be shown that mod σ is isomorphic to $\bar{\sigma}_W$ [3] .

If the associated flow $\{\tilde{T}_s\}_{s\in R}$ is finite measure (ν) preserving then a subspace $\{ f(w,w') \in L^2(W\ W,\nu\times\nu)$; for any s $f(\tilde{T}_s w,\tilde{T}_{-s}w') = f(w,w')$ a.e.(w,w') } of $L^2(W\ W,\nu\times\nu)$ is spanned by functions $\xi_t(w)\xi_t(w')$, where $\xi_t(w)$ is an eigen function of the flow $\{\tilde{T}_s\}_{s\in R}$ corresponding to an eigen value t [3] . Therefore $\bar{\sigma}_W$ is the identity map and so is mod σ.

Connes and Woods [5] proved that the module of Sakai's flip of any infinite direct product group of automorphisms is the identity map. Here we give two different proofs of this fact. We are given an infinite direct product measure space $(\prod_{n=1}^{\infty}\{0,1,\ldots,k_n-1\}, \prod_{n=1}^{\infty} P_n)$, where P_n is a probability measure on $\{0,1,\ldots,k_n-1\}$. The infinite direct product group of automorphism of the space is the group G generated by automorphisms g_n, $n=1,2,\ldots$, defined by $(g_n\omega)_n = \omega_n+1 \pmod{k_n}$ and $(g_n\omega)_i = \omega_i$ for $i\neq n$.

Set $\sigma_n(\omega_1,\omega_2,\ldots,\omega_n,\omega_{n+1},\ldots,\omega_1',\omega_2',\ldots,\omega_n',\omega_{n+1}',\ldots)$
$= (\omega_1',\omega_2',\ldots,\omega_n',\omega_{n+1},\ldots,\omega_1,\omega_2,\ldots,\omega_n,\omega'_{n+1},\ldots)$,
then σ_n is an automorphism in $[G\times G]$. Since σ_n is measure preserving, σ_n converges to σ as $n\to\infty$. By the theorem in the previous section mod σ is the identity map.

The other proof is the following. We show that every bounded $\tilde{G}$-invariant measurable function is given by a limit function $\lim_{n\to\infty} f_n(u+\sum_{i=1}^{n}\log P_i(\omega_i))$, where f_n's are measurable functions. From this fact it is easy to see that mod σ is the identity map. Let $f(\omega,u)$ be a bounded $\tilde{G}$-invariant measurable function. Set $f_n(\omega,u) = E(f(\cdot,u)/\bigvee_{i=1}^{n} F_i)$, where F_i is the smallest σ-algebra which makes the i-th coordinate measurable. Since $\log\frac{dPg}{dP}(\omega) = \sum_{i=1}^{n}(\log P_i(g\omega)-\log P_i(\omega))$ for g in the subgroup generated by $g_1, g_2,\ldots,g_n$, we have for such a g $f_n(g\omega,u-\sum_{i=1}^{n}\log P_i(g\omega)) = f_n(\omega,u-\sum_{i=1}^{n}\log P_i(\omega))$. Therefore $f_n(\omega,u-\sum_{i=1}^{n}\log P_i(\omega))$ does not depend on ω for a.e. (ω,u) and then we denote it by $f_n(u)$. By the martingale convergence theorem we have $\lim_{n\to\infty} f_n(u+\sum_{i=1}^{n}\log P_i(\omega)) = f(\omega,u)$ a.e. (ω,u).

REFERENCES

1. CONNES, A., and KRIEGER, W. Measure space automorphisms, the normalizers of their full groups, and approximate finiteness. (preprint).
2. CONNES, A., and TAKESAKI, M. The flow of weights on factors of type III . Tohoku Math. J. 29(1977), 473-575.
3. HAMACHI, T., and OSIKAWA, M. Ergodic groups of automorphisms and Krieger's Theorems. (To appear).
4. KRIEGER, W. On ergodic flows and isomorphism of factors. Math. Ann. 223 (1976), 19-70.
5. WOODS, E. J. A construction of approximately finite dimensional non-ITPFI factors. Symposium on operator algebra theory and its applications, Kyoto, 1977.

T. Hamachi and M.Oshikawa
Department of Mathematics
Ropponmatsu
4-2-1 Fukuoka
810 Japan

SOME REMARKS ON ε-INDEPENDENCE OF PARTITIONS AND ON TOPOLOGICAL ROCHLIN SETS

Gilbert Helmberg

I

Let $P = \{P_1,\dots,P_n\}$ and $Q = \{Q_1,\dots Q_m\}$ be measurable partitions of a probability space (X,F,μ). We shall use the following notation:

$$x_i = \mu(P_i), \quad y_{ij} = \mu(P_i \cap Q_j)/\mu(Q_j), \quad z_j = \mu(Q_j);$$

$$\eta(x) = \begin{cases} -x \log x & \text{for } 0 < x \leq 1, \\ 0 & \text{for } x = 0; \end{cases}$$

$$h(P) = \sum_{i=1}^{n} \eta(x_i), \qquad h(P/Q) = \sum_{j=1}^{m} z_j \sum_{i=1}^{n} \eta(y_{ij}).$$

Without loss of generality we assume all x_i and z_j to be positive and $0 < x_1 \leq x_2 \leq \dots \leq x_n < 1$.

Let $0 < \varepsilon \leq \frac{2}{e}$ (for reasons of convenience). According to Ornstein [5][6] P is called ε-independent of Q ($P \overset{\varepsilon}{\perp} Q$) if for some index set $J \subset \{1,\dots,m\}$ one has

$$\begin{cases} \sum_{i=1}^{n} |x_i - y_{ij}| < \varepsilon & \text{for all } j \in J, \\ \sum_{j \notin J} z_j < \varepsilon. \end{cases}$$

ε-independence is controlled by the quantity

$$h(P)-h(P/Q) = \sum_{j=1}^{m} z_j \sum_{i=1}^{n} (\eta(x_i)-\eta(y_{ij}))$$

[8, thm.4.8]: if $h(P)-h(P/Q)$ is small (e.g. smaller than $\frac{\varepsilon^4}{54} \log \frac{e}{2}$ [2, Satz 3]) then $P \overset{\varepsilon}{\perp} Q$. Still one may have $P \overset{\varepsilon}{\perp} Q$ without $h(P)-h(P/Q)$ being small, due to a large number of atoms of P: for $x_1 = \dots = x_{n-1} = \frac{\varepsilon}{2(n-1)}$, $x_n = 1 - \frac{\varepsilon}{2}$ one has $P \overset{\varepsilon}{\perp} Q$ but

$$h(P)-h(P/P) = h(P) = \frac{\varepsilon}{2}\log(n-1)+\eta(\tfrac{\varepsilon}{2})+\eta(1-\tfrac{\varepsilon}{2})$$

[7]. As noticed by Fleischmann [2] and the author [3] this assertion may be made more precise as follows.

Theorem: Let n denote the number of non-zero atoms of P.

a) For every pair of partitions P and Q satisfying $P \overset{\varepsilon}{\perp} Q$ one has

$$h(P)-h(P/Q) \le \frac{3-\varepsilon}{2}\,\varepsilon\log n + (1-\varepsilon)(\eta(\tfrac{\varepsilon}{2})+\eta(1-\tfrac{\varepsilon}{2})).$$

b) There exist partitions P and Q (depending on ε and n) satisfying $P \overset{\varepsilon}{\perp} Q$ and

$$h(P)-h(P/Q) = \frac{3-\varepsilon}{2}\,\varepsilon\log n - (1-\varepsilon)\eta(\tfrac{\varepsilon}{2}) + O(\tfrac{\log n}{n}) \quad \text{as } n \to \infty.$$

It seems convenient to prepare the proof by a series of auxiliary assertions involving the concavity of the function η.

a) $\left.\begin{array}{l} 0 \le x < u \le v < y \le 1 \\ x+y = u+v \end{array}\right\} \Rightarrow \eta(x)+\eta(y) < \eta(u)+\eta(v)$

(by the strict concavity of η).

b) $\left.\begin{array}{l} 0 < u \le v < 1 \\ u+v \le 1 \end{array}\right\} \Rightarrow \eta(u+v) < \eta(u)+\eta(v)$

(put $x = 0$, $y = u+v$ in a)).

c) Suppose $\begin{array}{l} 0 < y_i \le x_i < 1, \\ 0 < x_j < y_j < 1, \\ x_j < x_i. \end{array}$

Then there exists a $\tau > 0$ such that

$$\eta(x_i)-\eta(y_i)+\eta(x_j)-\eta(y_j) < \eta(x_i)-\eta(y_i+\tau)+\eta(x_j)-\eta(y_j-\tau),$$

$$|x_i-y_i| + |x_j-y_j| \ge |x_i-y_i-\tau| + |x_j-y_j+\tau|.$$

In fact, choose any positive $\tau \le \min(y_j, 1-y_i)$ satisfying

$$y_j-y_i < \tau \le y_j-y_i+(x_i-x_j) = |x_i-y_i| + |x_j-y_j|.$$

The first inequality then follows from a), the second one may be checked directly.

d) $\left.\begin{array}{l} 0 < x_i < y_i \le 1 \\ 0 < x_j < y_j \le 1 \\ x_i \le x_j \end{array}\right\} \Rightarrow \eta(x_i)-\eta(y_i)+\eta(x_j)-\eta(y_j) < \eta(x_j)-\eta(y_j+y_i-x_i)$

(put $x = x_i$, $y = y_j+y_i-x_i$ in a)).

e) $\left.\begin{array}{l} 0 < x_j < x_i \\ 0 \le y_i < x_i \end{array}\right\} \Rightarrow \eta(x_i)-\eta(y_i) < \begin{cases} \eta(x_j)-\eta(x_j-x_i+y_i) & \text{if } x_j-x_i+y_i \ge 0, \\ \eta(x_j)+\eta(x_i)-\eta(y_i+x_j) & \text{if } x_j-x_i+y_i < 0. \end{cases}$

For the first inequality, put $x = x_j-x_i+y_i$, $y = x_i$ in a); the second inequality follows from b).

f) $\left.\begin{array}{l} 0 < y_i < x_i \\ 0 < y_j < x_j \\ x_i \le x_j \end{array}\right\} \Rightarrow \eta(x_i)-\eta(y_i)+\eta(x_j)-\eta(y_j) < \begin{cases} \eta(x_i)+\eta(x_j)-\eta(y_i+y_j) \\ \quad \text{if } y_i+y_j \le x_j, \\ \eta(x_i)-\eta(y_i+y_j-x_j) \\ \quad \text{if } y_i+y_j > x_j. \end{cases}$

The first inequality follows from b), for the second one put $x = y_i+y_j-x_j$, $y = x_j$ in a).

g) $\left.\begin{array}{l} 0 < y_i \le x_i \le x_j \le y_j < 1 \\ 0 < \tau \le \min\,(y_i, 1-y_j) \end{array}\right\} \Rightarrow \begin{array}{l} \eta(x_i)-\eta(y_i)+\eta(x_j)-\eta(y_j) < \\ < \eta(x_i)-\eta(y_i-\tau)+\eta(x_j)-\eta(y_j+\tau) \end{array}$

(put $x = y_i-\tau$, $y = y_j+\tau$ in a)).

<u>Lemma 1</u>: Let $E(x_1,\ldots,x_n;\varepsilon)$ denote the maximum of the convex function $\sum_{i=1}^{n} (\eta(x_i)-\eta(y_i))$ in $(y_1,\ldots y_n)$ in the closed convex subset A of R^n defined by

$$\begin{cases} 0 \le y_i, & (1 \le i \le n), \\ \sum_{i=1}^{n} y_i = 1, & \\ \sum_{i=1}^{n} |x_i-y_i| \le \varepsilon. & \end{cases}$$

Let α and k be defined by

$$\alpha = \min\,(1-x_n\,, \tfrac{\varepsilon}{2})$$

$$\sum_{i=1}^{k-1} x_i < \alpha \le \sum_{i=1}^{k} x_i.$$

Then $E(x_1,\ldots,x_n;\varepsilon) = \sum_{i=1}^{k} \eta(x_i)-\eta(\sum_{i=1}^{k} x_i-\alpha)+\eta(x_n)-\eta(x_n+\alpha)$

<u>Proof</u>: Suppose the maximum is attained in the point $(\overline{y_1},\ldots,\overline{y_n}) \in A$ and let

$$\beta = \sum_{i=1}^{n} |x_i - \overline{y_i}| \leq \varepsilon,$$

$$\sum_{\overline{y_i}<x_i} (x_i - \overline{y_i}) = \sum_{x_i<\overline{y_i}} (\overline{y_i} - x_i) = \frac{\beta}{2}.$$

Note that $\overline{y}_i = 0 < \overline{y}_j$ and $\overline{y}_i < \overline{y}_j = 1$ (if necessary after an interchange of $\overline{y}_i$ and $\overline{y}_j$) allows to assume without loss of generality $x_i \leq x_j$.

The following steps 1) - 5) serve to complete the proof by showing

$$\begin{aligned} \overline{y_n} &= x_n + \alpha, & & \\ \overline{y_j} &= x_j & & \text{for } k+1 \leq j \leq n-1, \\ \overline{y_h} &= \sum_{i=1}^{k} x_i - \alpha & & \text{for some } h,\ 1 \leq h \leq k, \\ \overline{y_i} &= 0 & & \text{otherwise.} \end{aligned}$$

The reasoning in every step is that in the case contrary to the assertion one could choose a point $(y_1, \dots y_n) \in A$ satisfying

$$\sum_{i=1}^{n} (\eta(x_i) - \eta(y_i)) > \sum_{i=1}^{n} (\eta(x_i) - \eta(\overline{y_i})).$$

1) For $\overline{y_i} \leq x_i$ and $x_j < \overline{y_j}$ one has $x_i \leq x_j$ by c).

2) The inequality $x_j < \overline{y_j}$ can hold for at most one index j by d). By 1) one then has $j = n$ and therefore (in any case) $\overline{y_n} - x_n = \frac{\beta}{2}$.

3) For $\overline{y_i} < x_i$ and $x_j = \overline{y_j}$ one has $x_i \leq x_j$ by e). Consequently for some index $k < n$ one has

$$\begin{aligned} \overline{y_i} &< x_i & & \text{for } 1 \leq i \leq k, \\ \overline{y_j} &= x_j & & \text{for } k+1 \leq j \leq n-1. \end{aligned}$$

4) The inequality $0 < \overline{y_h} < x_h$ can hold for at most one index $h \leq k$ by f). This implies

$$\sum_{i=1}^{k-1} x_i < \frac{\beta}{2} \leq \sum_{i=1}^{k} x_i.$$

5) One has $\beta > 0$ by g). If $\beta = \varepsilon$, then $\frac{\varepsilon}{2} = \frac{\beta}{2} = \overline{y_n} - x_n \leq 1 - x_n$. If $\beta < \varepsilon$, then g) implies $\overline{y_n} = x_n + \frac{\beta}{2} = 1$. Consequently

$$\alpha = \frac{\beta}{2} = \min\left(\frac{\varepsilon}{2}, 1 - x_n\right).$$

Lemma 2. $E(x_1,\dots,x_n;\varepsilon) \le E(n,\varepsilon) = \eta(\frac{\varepsilon}{2})+\eta(1-\frac{\varepsilon}{2})+\frac{\varepsilon}{2}\log(n-1)$.

Proof. Let $k < n$ and $\alpha \le \frac{\varepsilon}{2}$ be as in lemma 1. Using b) we get

$$\eta(x_k)-\eta(\sum_{i=1}^{k} x_i-\alpha) \le \eta(\alpha-\sum_{i=1}^{k-1} x_i),$$

$$\sum_{i=1}^{k}\eta(x_i)-\eta(\sum_{i=1}^{k} x_i-\alpha) \le \sum_{i=1}^{k-1}\eta(x_i)+\eta(\alpha-\sum_{i=1}^{k-1} x_i) \le k\eta(\frac{\alpha}{k}) =$$

$$= \eta(\alpha)+\alpha\log k \le \eta(\frac{\varepsilon}{2}) + \frac{\varepsilon}{2}\log(n-1)$$

and putting $x = x_n$, $y = 1$ in a)

$$\eta(x_n)-\eta(x_n+\alpha) \le \eta(1-\alpha) \le \eta(1-\frac{\varepsilon}{2}).$$

Proof of the theorem.

a) Let $\sum_{j\notin J} z_j = \gamma < \varepsilon$. Then by lemma 1 and lemma 2 we have

$$\sum_{j=1}^{m} z_j \sum_{i=1}^{n} (\eta(x_i)-\eta(y_{ij})) \le (1-\gamma)E(n,\varepsilon)+\gamma\log n < (1-\varepsilon)E(n,\varepsilon)+\varepsilon\log n.$$

The last inequality holds since

$$E(n,\varepsilon) \le E(n,\frac{2}{e}) < \log n$$

b) For convenience suppose ε to be irrational. Let $k = [\frac{n\varepsilon}{2}]$ and choose $P \vee Q$ such that $m = 2n$ and

$$\mu(P_i\cap Q_j) = \begin{cases} \frac{k+1}{n^2}(1-\frac{2k}{n}) & \text{for } 1 \le i = j \le n \\ \frac{1}{n^2}(1-\frac{2k}{n}) & \text{for } 1 \le j \le n,\ i \ne j-l \pmod n,\ 0 \le l \le k \\ 0 & \text{for } 1 \le j \le n,\ i = j-l \pmod n,\ 1 \le l \le k \\ \frac{2k}{nn^2} & \text{for } n+1 \le j \le 2n,\ i = j-n \\ 0 & \text{for } n+1 \le j \le 2n,\ i \ne j-n. \end{cases}$$

Then $$x_i = \sum_{j=1}^{2n}\mu(P_i\cap Q_j) = \frac{1}{n}$$

$$z_j = \sum_{i=1}^{n}\mu(P_i\cap Q_j) = \begin{cases} \frac{1}{n}(1-\frac{2k}{n}) & \text{for } 1 \le j \le n \\ \frac{2k}{n^2} & \text{for } n+1 \le j \le 2n. \end{cases}$$

The relation $P \overset{\varepsilon}{|} Q$ follows from

$$\sum_{i=1}^{n} |x_i - y_{ij}| = \left(\frac{k+1}{n} - \frac{1}{n}\right) + \frac{k}{n} = \frac{2k}{n} < \varepsilon \qquad \text{for } 1 \leq j \leq n,$$

$$\sum_{j=n+1}^{2n} z_j = \frac{2k}{n} < \varepsilon.$$

A straight-forward computation furnishes

$$h(P)-h(P/Q) = \left(1-\frac{2k}{n}\right)\frac{k+1}{n} \log (k+1) + \frac{2k}{n} \log n$$

$$= \frac{3-\varepsilon}{2} \varepsilon \log n - (1-\varepsilon)\eta\left(\frac{\varepsilon}{2}\right) + O\left(\frac{\log n}{n}\right) \quad \text{as } n \to \infty.$$

II.

Let T be an aperiodic homeomorphism of a compact metric space X. In [1] (15.6) the existence of a topological generator is proved using a lemma (15.5) which in turn is based on a proposition (15.4) interesting in its own right since it asserts the existence of topological analoga of Rochlin sets. As proved in [1] it may be stated as follows:

Given any subset A of X and any $n \in \mathbb{N}$ there exists a subset V of A such that

1) $V \subset \overline{\overset{\circ}{V}}^{(A)}$,

2) $V \cap T^j V = \emptyset$ for $1 \leq j \leq n-1$,

3) $A \subset \bigcup_{k=-n+1}^{n-1} T^k \overline{V}$,

4) $\overline{V}^{(A)} \subset \bigcup_{k=-n+1}^{n-1} T^k V$.

Here $\overline{V}$ denotes the closure of V in X while $\overline{V}^{(A)}$ resp. $\overline{\overset{\circ}{V}}^{(A)}$ denotes closure resp. closure of the interior of V with respect to the relative topology in A.

It is tempting to try to replace $\overline{V}$ by $\overline{V}^{(A)}$ in 3) or $\overline{V}^{(A)}$ by $\overline{V}$ in 4), both resulting in $A \subset \bigcup_{k=-2(n-1)}^{2(n-1)} T^k V$ (as happened in [1] due to a mixup of both topologies). However, the following example [4] (with n = 2) shows that this is impossible in general:

Let $\alpha \in]0,\frac{1}{4}[$ be irrational and define T on $\mathbb{R}/\mathbb{Z}$ by

$Tx \equiv x + \alpha \quad (\text{mod } 1).$

Let $A =]0,\alpha[\cup (]\alpha,2\alpha]\setminus\mathbb{Q})$. If a subset V of A satisfies $V \subset \overset{\circ}{\overline{V}}{}^{(A)}$ and $V \cap TV = \emptyset$, then $B = A \setminus \bigcup_{k=-2}^{2} T^k\overline{V}^{(A)} \neq \emptyset$. Indeed, it may readily be checked that if $2\alpha \notin \overline{V}^{(A)}$, then $2\alpha \in B$, while if $2\alpha \in \overline{V}^{(A)}$, then $[\alpha-\varepsilon,\alpha-\eta[\cap T^{-1}\mathbb{Q} \subset B$ for some $\varepsilon > \eta > 0$.

A reexamination of the proof of (15.5) in [1] reveals that it may be carried through also using the corrected version of proposition (15.4) as formulated above. Lemma (15.5) may even slightly be sharpened (as compared with the formulation in [1]) putting $c_n = 2\sum_{k=1}^{n} p_k$ and requiring $X_\alpha \subset \bigcup_{k=-c_n}^{c_n} T^k \overset{\circ}{A}_n$ for all $n \in \mathbb{N}$.

REFERENCES.

[1] Denker, M./Grillenberger, C./Sigmund, K.: Ergodic Theory in Compact Spaces. Lecture Notes in Mathematics 527. Berlin-Heidelberg-New York: Springer Verlag 1976.

[2] Fleischmann, E.: Über ε-Unabhängigkeit von Partitionen. Diss. Phil. Fak. Universität Innsbruck 1977.

[3] Helmberg, G.: Über ε-Unabhängigkeit von Partitionen. Seminarnotiz Nr.4, Nr.6, Institut für Mathematik I, Techn. Fak. Universität Innsbruck 1975/76.

[4] Helmberg, G./Wagner P.: Über topologische Analoga von Rochlin's Lemma. Seminarnotiz Nr.3, Institut für Mathematik I, Techn. Fak. Universität Innsbruck 1977/78.

[5] Ornstein, D.S.: Bernoully shifts with the same entropy are isomorphic. Advances in Math. 4, 337-352 (1970).

[6] Ornstein, D.S.: Ergodic Theory, Randomness, and Dynamical Systems. Yale Mathematical Monographs 5. New Haven-London: Yale University Press 1974.

[7] Roider, B.: Über ε-Unabhängigkeit von Partitionen. Seminarnotiz Nr.2, Institut für Mathematik I, Techn. Fak. Universität Innsbruck 1975/76.

[8] Smorodinsky, M.: Ergodic Theory, Entropy. Lecture Notes in Mathematics 214. Berlin-Heidelberg-New York: Springer Verlag 1971.

Gilbert Helmberg
Universität Innsbruck
1. Lehrkanzel f. Mathematik
Technikerstr. 13
A-6020 Innsbruck

Maximal measures for piecewise monotonically increasing transformations on [0,1]

Franz Hofbauer

Introduction

In chapter I we consider dynamical systems (I,f), where $I = [0,1] = \bigcup_{i=1}^{n} J_i$, J_i disjoint intervals and $f|J_i$ continuous and strictly increasing. Furthermore we assume

(a) $(J_1, J_2, \ldots, J_n)$ is a generator for (I,f)
(b) $h_{top}(f) > 0$

Our goal is to determine the set of invariant measures μ on (I,f) with maximal entropy, i.e. the entropy $h(\mu)$ of μ is equal to the topological entropy $h_{top}(f)$ of (I,f) by the variational principle.

The f-expansion gives an isomorphism φ of (I,f) with a subshift Σ_f^+ of $\Sigma_n^+ = \{1,2,\ldots,n\}^{\mathbb{N}}$. We construct a two-sided subshift Σ_M of finite type over an infinite alphabet, M the corresponding infinite transition matrix, which is in some sense isomorphic to Σ_f, the natural extension of Σ_f^+, such that the set of maximal measures is preserved. If $M^1, M^2, \ldots$ are the irreducible submatrices of M, we have the following results.

(i) $h_{top}(\Sigma_M) = \log r(M)$, the logarithm of the spectral radius of M
(ii) every ergodic maximal measure is concentrated on a Σ_{M^i} satisfying $r(M^i) = r(M)$
(iii) for every such M^i there is at most one maximal measure concentrated on Σ_{M^i}. It is Markov.

In chapter II we consider the transformations $T: x \mapsto \beta x + \alpha \pmod 1$ for $0 \le \alpha < 1$ and $\beta > 1$, which are the most important examples for the transformations of chapter I. We compute M in this case and we have the following results.

(i) $h_{top}(T) = \log \beta$

(ii) T has unique maximal measure μ

(iii) μ is absolutely continuous with respect to Lebesgue measure and its support is a finite union of intervals.

Chapter I

1. Let (I,f) be as in the introduction. Σ_n^+ denotes the full one-sided shift over $\{1,2,\dots,n\}$, σ the shift-transformation and $\leq$ the lexicographic ordering on Σ_n^+. The f-expansion $\varphi: I \to \Sigma_n^+$ defined by $\varphi(x) = \underline{x} = x_0 x_1 x_2 \dots$, such that $f^i(x) \in J_{x_i}$ for $i \geq 0$, is injective because of (a) and order preserving, because $f|J_i$ is increasing. Define $\Sigma_f^+ = \overline{\varphi(I)} \subset \Sigma_n^+$. $\Sigma_f^+ \setminus \varphi(I)$ is countable, hence a nullset for every maximal measure and therefore the set of maximal measures is preserved by φ. If $J_i = (r,s)$ set $\underline{a}^i = \lim_{t \downarrow r} \varphi(t)$ and $\underline{b}^i = \lim_{t \uparrow s} \varphi(t)$. Then we have the following generalization of the β-shift.

<u>Lemma 1.</u> $\Sigma_f^+ = \{\underline{x} \in \Sigma_n^+ : \underline{a}^{x_i} \leq \sigma^i \underline{x} \leq \underline{b}^{x_i} \ \forall\ i \geq 0\}$.

<u>Proof.</u> It is easy to see that Σ_f^+ is a subset of this set using the fact that φ is order preserving. For the other inclusion one has to prove that the intervals $J_{x_0} \cap f^{-1} J_{x_1} \cap \dots \cap f^{-i} J_{x_i}$ are not empty for $i \geq 1$, if $\underline{a}^{x_i} \leq \sigma^i \underline{x} \leq \underline{b}^{x_i}$ for $i \geq 0$. This can be done by induction.

We shall need the following lemma later.

<u>Lemma 2.</u> Let $\underline{x} \in \Sigma_f^+$ and $\underline{a} = \underline{a}^k$ (or $\underline{b}^k$), $\underline{b} = \underline{a}^m$ (or $\underline{b}^m$) for some k and m. If

$x_j x_{j+1} \dots x_{j+r} = a_0 a_1 \dots a_r$ and

$x_{j+r} x_{j+r+1} \dots x_{j+r+s} = b_0 b_1 \dots b_s$ then

$x_{j+r} x_{j+r+1} \dots x_{j+r+s} = b_0 b_1 \dots b_s = a_r a_{r+1} \dots a_{r+s}$.

<u>Proof.</u> Suppose this is not satisfied, i.e. there is an i $(0 \leq i \leq s-1)$ such that $x_{j+r} \dots x_{j+r+i} = b_0 \dots b_i = a_r \dots a_{r+i}$ and $x_{j+r+i+1} = b_{i+1} \neq a_{r+i+1}$. But $b_{i+1} < a_{r+i+1}$ implies $\sigma^j \underline{x} < \underline{a}$, a

contradiction to lemma 1 and $b_{i+1} > a_{r+i+1}$ implies $\sigma^r \underline{a} < \underline{b}$, again a contradiction to lemma 1.

2. Σ_f^+ can be characterized by all blocks $x_0 x_1 \dots x_{m-1}$ which are admissible in Σ_f^+, i.e. ${}_0[x_0 x_1 \dots x_{m-1}] = \{\underline{z} \in \Sigma_f^+ : z_0 = x_0, z_1 = x_1, \dots, z_{m-1} = x_{m-1}\}$ is not empty. This is equivalent to $\sigma^m\, {}_0[x_0 x_1 \dots x_{m-1}] \neq \emptyset$. Set $G_{x_0 \dots x_{m-1}} = \sigma^m\, {}_0[x_0 \dots x_{m-1}]$.
We have $G_{x_0} = \sigma_0[x_0] = \sigma([\underline{a}^{x_0}, \underline{b}^{x_0}]) = [\sigma\underline{a}^{x_0}, \sigma\underline{b}^{x_0}]$, a closed interval in Σ_f^+. Suppose $G_{x_0 \dots x_{m-1}} = [\sigma^k \underline{a}^i, \sigma^l \underline{b}^j]$. Then

$$
\begin{aligned}
G_{x_0 \dots x_m} &= \sigma^{m+1}\, {}_0[x_0 \dots x_m] = \sigma({}_0[x_m] \cap \sigma^m\, {}_0[x_0 \dots x_{m-1}]) \\
&= \sigma\,([\underline{a}^{x_m}, \underline{b}^{x_m}] \cap G_{x_0 \dots x_{m-1}}) \\
&= \sigma\,([\underline{a}^{x_m}, \underline{b}^{x_m}] \cap [\sigma^k \underline{a}^i, \sigma^l \underline{b}^j]) \\
&= \begin{cases}
\emptyset \text{ if } x_m < a_{k+1} \text{ or } x_m > b_{l+1} \\
[\sigma^{k+1}\underline{a}^i, \sigma\underline{b}^{x_m}] \text{ if } x_m = a_{k+1} \text{ and } x_m < b_{l+1} \\
[\sigma\underline{a}^{x_m}, \sigma^{l+1}\underline{b}^j] \text{ if } x_m > a_{k+1} \text{ and } x_m = b_{l+1} \\
[\sigma^{k+1}\underline{a}^i, \sigma^{l+1}\underline{b}^j] \text{ if } x_m = a_{k+1} = b_{l+1} \\
[\sigma\underline{a}^{x_m}, \sigma\underline{b}^{x_m}] \text{ if } a_{k+1} < x_m < b_{l+1}
\end{cases}
\end{aligned}
$$

Hence $G_{x_0 \dots x_{m-1}}$ is a closed interval in Σ_f^+ or empty.
$G_{x_0 \dots x_{m-1}} = \emptyset$, iff $x_0 \dots x_{m-1}$ is not admissible in Σ_f^+.

We may form a diagram with the $G_{x_0 \dots x_{m-1}}$ (we take $n = 2$ for convenience). It will be called M.

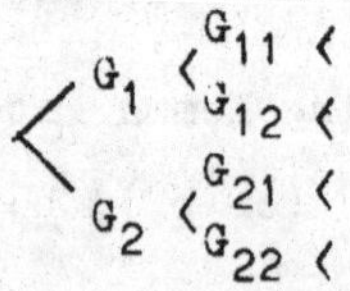

Every path in this diagram leading not to an empty set represents an $\underline{x} \in \Sigma_f^+$ and every $\underline{x} \in \Sigma_f^+$ is represented by such a path. We can represent $\underline{x}$ also by the $G_{x_0 \dots x_{m-1}}$ on its path.
Define $\psi' : \Sigma_f^+ \to D^{\mathbb{N}}$, where $D = \{(x_{m-1}, G_{x_0 \dots x_{m-1}}) : G_{x_0 \dots x_{m-1}} \neq \emptyset\}$

$= \{(a^i_{k-1}, [\sigma^k \underline{a}^i, \sigma^l \underline{b}^j]): k,l \geq 1 \text{ and } a^i_{k-m} = b^j_{l-m} \text{ for } 1 \leq m \leq \min(k,l)\}$

by

$$\psi'(\underline{x}) = (x_0, G_{x_0})(x_1, G_{x_0 x_1})\ldots$$

Unfortunately $\psi'(\Sigma_f^+)$ is not stable under σ. So we have to use the natural extension $\Sigma_f = \{\underline{x} \in \Sigma_n = \{1,\ldots,n\}^{\mathbf{Z}}: x_i x_{i+1} \ldots \in \Sigma_f^+ \ \forall\ i \in \mathbf{Z}\}$. Let N be the following σ-invariant subset of Σ_f. $N = \{\underline{x} \in \Sigma_f: \exists\ m \in \mathbf{Z}$ so that $\forall\ j < m\ \ \exists\ k \leq j$ with $x_k x_{k+1} \ldots x_m = a^i_0 \ldots a^i_{m-k}$ or $b^i_0 \ldots b^i_{m-k}$ for some i and $\ldots x_{m-2} x_{m-1} x_m$ is not periodic$\}$.

Now let $\underline{x} \in \Sigma_f$ and define for each $k \leq 0$ $\underline{y}^k = \psi'(x_k x_{k+1} \ldots) = (x_k, G_{x_k})(x_{k+1}, G_{x_k x_{k+1}})\ldots \in \prod_k^\infty D$, i.e. we have

$$\begin{array}{l} y^0_0\, y^0_1\, y^0_2 \ldots = \underline{y}^0 \\ y^{-1}_{-1} y^{-1}_0\, y^{-1}_1\, y^{-1}_2 \ldots = \underline{y}^{-1} \\ y^{-2}_{-2} y^{-2}_{-1} y^{-2}_0\, y^{-2}_1\, y^{-2}_2 \ldots = \underline{y}^{-2} \\ \ldots\ldots\ldots\ldots\ldots\ldots \end{array}$$

<u>Lemma 3.</u> For $\underline{x} \in \Sigma_f \setminus N$ there is a $\underline{y} \in D^{\mathbf{Z}}$ with $\underline{y}^k \to \underline{y}$ $(k \to -\infty)$, i.e. the columns in the above table are finally constant.

<u>Proof.</u> Fix m. Because $\underline{x} \notin N$ there is a j such that for every $k \leq j$ $x_k x_{k+1} \ldots x_m$ is no initial segment of an $\underline{a}^i$ or $\underline{b}^i$ or $\ldots\ x_{m-1} x_m$ is periodic. In the first case let $r \leq j$ and s be so that $x_r x_{r+1} \ldots x_s$ is an initial segment of some $\underline{a}^i$ and has largest possible s. We have $s < m$. By lemma 2 every initial segment of an $\underline{a}^i$ in $\underline{x}$ which begins left of or at the s-th coordinate ends also left of or at the s-th coordinate. The same can be done for the $\underline{b}^i$'s. There is a $t < m$ such that every initial segment of a $\underline{b}^i$ in $\underline{x}$ which begins left of or at the t-th coordinate ends also left of or at the t-th coordinate. Hence one has for $k \leq j$

$$\underline{y}^k = (x_k, [\sigma \underline{a}^{x_k}, \sigma \underline{b}^{x_k}]) \ldots (x_{s+1}, [\sigma \underline{a}^{x_{s+1}}, \ ]) \ldots (x_{t+1}, [\ , \sigma \underline{b}^{x_{t+1}}])\ldots$$

Because of $s < m$ and $t < m$ all $\underline{y}^k$ for $k \leq j$ have the same entries from the m-th coordinate onwards. If $\ldots x_{m-2} x_{m-1} x_m$ is periodic and there is a periodic $\underline{a}^i$ with the same period then we use the fact that $\sigma^{p+m}\underline{a}^i = \sigma^m\underline{a}^i$ $\forall$ m for some p to get the desired result.

Now define ($\underline{y}$ as in lemma 3)

$$\psi: \Sigma_f \setminus N \to D^{\mathbf{Z}} \text{ by } \psi(\underline{x}) = \underline{y}$$

By the computation at the beginning of § 2 one sees easily that a given element $(a^i_{k-1}, [\sigma^k\underline{a}^i, \sigma^l\underline{b}^j])$ of D can be followed in a point $\underline{y} \in \psi(\Sigma_f\setminus N)$ by (we write r for a^i_k and s for b^j_l)

$(r, [\sigma^{k+1}\underline{a}^i, \sigma\underline{b}^r), (s, [\sigma\underline{a}^s, \sigma^{l+1}\underline{b}^j])$ and

$(m, [\sigma\underline{a}^m, \sigma\underline{b}^m])$ for $r < m < s$, if $r < s$

$(r, [\sigma^{k+1}\underline{a}^i, \sigma^{l+1}\underline{b}^j])$ if $r = s$

Denote the corresponding transition matrix with index set D and entries 0's and 1's by M (it is that one given by the diagram of the $G_{x_0 \dots x_{m-1}}$). Define

$$\Sigma_M = \{\underline{y} \in D^{\mathbf{Z}} : M_{y_i y_{i+1}} = 1 \ \forall \ i \in \mathbf{Z}\}$$

Then $\psi(\Sigma_f\setminus N) \subset \Sigma_M$.

ψ^{-1} is given by the projection to the first component of the elements of D. We have

<u>Lemma 4.</u> Let $\underline{y} \in \Sigma_M$. Then $\underline{x} = \psi^{-1}(\underline{y}) \in \Sigma_f$

<u>Proof.</u> We have to show that $x_k x_{k+1} \dots \in \Sigma_f^+$ for all k. By definition of ψ, $x_k x_{k+1} \dots$ is a path of arrows in M beginning at some element of D. But there is a path $z_1 z_2 \dots z_m$ from the beginning of the diagram M (i.e. z_1 is an arrow leading to some G_i, $1 \le i \le n$) to this element. Therefore $z_1 \dots z_m x_k x_{k+1} \dots \in \Sigma_f^+$ and hence also $x_k x_{k+1} \dots =$
$= \sigma^m (z_1 \dots z_m x_k x_{k+1} \dots) \in \Sigma_f^+$.

It is easy to see that ψ and ψ^{-1} are measurable and that ψ commutes with the shift transformation. Hence ψ is an isomorphism between $(\Sigma_f\setminus N, \sigma)$ and (Σ_M, σ).

Lemma 5. If ν is a σ-invariant measure concentrated on N then $h(\nu)=0$.

$N=\bigcup N_{\underline{a}^i}\cup\bigcup N_{\underline{b}^i}$, the unions taken over all i and j such that $\underline{a}^i$ and $\underline{b}^i$ are not periodic, where $N_{\underline{x}}=\bigcup_{j\in\mathbb{Z}}\sigma^j\bigcap_{l=1}^{\infty}\bigcup_{k=1}^{\infty}\sigma^{-k}({}_0[x_0x_1\ldots x_{k-1}])$. It suffices to prove lemma 5 for $N_{\underline{x}}$, $\underline{x}=\underline{a}^i$ or $\underline{b}^j$. This is done in [1] constructing imbeddings of $(N_{\underline{x}},\sigma)$ into subshifts S_k of $\{0,1,2\}^{\mathbb{Z}}$ for every $k\in\mathbb{N}$ such that $h_{top}(S_k)\to 0$ for $k\to\infty$.

Remark. Σ_f^+ and Σ_M can also be defined, if $f|J_i$ is decreasing for some i's. But the proof that the set N, where ψ cannot be defined, is small in the above sense does not work, because it uses lemma 2 and this lemma does not hold any more.

From lemma 5 it follows that N is a nullset for every maximal measure, because we have assumed that $h_{top}(f)>0$. Hence

Theorem 1. (Σ_f,σ) and (Σ_M,σ) have isomorphic sets of maximal measures.

3. To compute the topological entropy $h_{top}(\Sigma_f^+)=$ $=\lim\frac{1}{k}\log n_k$, where n_k is the number of admissible blocks of length k in Σ_f^+, we introduce the following sets of blocks

$\mathcal{N}_k^g=\{x_0x_1\ldots x_{k-1}:\ {}_0[x_0\ldots x_{k-1}]\neq\emptyset,\ \psi'(x_0\ldots x_{k-1})$ ends with $g\}$, where $k\geq 1$ and $g\in D$. Set $N_k^g=\operatorname{card}\mathcal{N}_k^g$ and $N_k=(N_k^g)_{g\in D}$. We obtain the set $\mathcal{N}_{k+1}^h$ adding an x_k to $x_0\ldots x_{k-1}\in\bigcup_g\mathcal{N}_k^g$, such that ${}_0[x_0\ldots x_k]\subset\Sigma_f^+$ is not empty and that $\psi'(x_0\ldots x_k)$ ends with h. This is possible for all blocks $x_0\ldots x_{k-1}\in\mathcal{N}_k^g$, if g may be followed by h in M, i.e. $M_{gh}=1$. Hence we have, as x_k has to be equal to the first component of h,

$N_{k+1}^h=\sum_{g\in D}N_k^gM_{gh}$, i.e. $N_{k+1}=N_kM$. From this one gets $N_k=N_1M^k$ and as $n_k=\|N_k\|_1=\sum_{g\in D}N_k^g$ (N_k has only finitely many entries $\neq 0$) we have $h_{top}(\Sigma_f^+)=\lim\frac{1}{k}\log\|N_1M^k\|_1=r(M)$, the spectral radius of the l^1-operator $u \to uM$. We have

$$h_{top}(f) = h_{top}(\Sigma_f^+) = h_{top}(\Sigma_f) = h_{top}(\Sigma_M) = \log r(M).$$

$h_{top}(\Sigma_M) = h_{top}(\Sigma_f)$ follows from the variational principle.

Now divide M into irreducible submatrices $M^1, M^2, \ldots$ and consider an ergodic maximal measure μ on Σ_M. Because of ergodicity, μ must be concentrated on Σ_{M^i}, the subshift of Σ_M corresponding to the irreducible submatrix M^i, for some i. From a theorem of Parry [cf. 4] it follows that μ must be a Markov measure. Now we use the same method as Takahashi in [4] for the β-shift and get that μ is a Markov measure given by $\pi_g = u_g v_g$ and $P_{gh} = M^i_{gh} v_h / \lambda v_g$ $(g,h \in D)$, where u is a left and v a right eigenvector of M^i for the eigenvalue $\lambda = r(M^i)$. Furthermore $\log \lambda = h(\mu) = h_{top}(\Sigma_M) = \log r(M)$. Now we can use a result of Krieger [3, theorem 1] for the irreducible matrix M^i and get that there is at most one maximal measure on Σ_{M^i}. Hence we have

<u>Theorem 2.</u>

(i) $h_{top}(\Sigma_M) = \log r(M)$

(ii) every ergodic maximal measure is concentrated on a Σ_{M^i} satisfying $r(M^i) = r(M)$

(iii) for every such M^i there is at most one maximal measure concentrated on Σ_{M^i}.

<u>Chapter II</u>

1. The most important examples for transformations considered in chapter I are the transformations T: $x \to \beta x + \alpha$ (mod 1) for $0 \le \alpha < 1$ and $\beta > 1$. We want to compute M in this case. Let $\underline{a} = \underline{a}^1$ be the expansion of 0 and $\underline{b} = \underline{b}^n$ be the expansion of 1, n so that $\beta + \alpha \le n < \beta + \alpha + 1$. Then $\underline{a}^i = i\underline{a}$ for $2 \le i \le n$ and $\underline{b}^i = i\underline{b}$ for $1 \le i \le n-1$. We have

$$\Sigma_T^+ = \{\underline{x} \in \Sigma_n^+ : a \le \sigma^k \underline{x} \le \underline{b} \ \forall\ k \ge 0\}.$$

Now divide $\underline{a}$ into initial segments of $i\underline{b}$ $(1 \le i \le n-1)$. $a_0 = 1$, hence there is an initial segment of $1\underline{b}$ at the beginnning of $\underline{a}$, which has at least length one. Denote its length by r_1 ($r_1 \geq 1$). Then $a_j = b_{j-1}$ for $1 \leq j \leq r_1 - 1$ and $a_{r_1} < b_{r_1-1} \leq n$. Hence there is the beginning of an initial segment of $a_{r_1}\underline{b}$ ($a_{r_1} < n$) of length at least one. Denote its length by r_2. We continue in this way and get $r_1, r_2, \ldots$ ($r_i \geq 1$) such that for $k \geq 0$

$$a_{r_1+\ldots+r_k+j} = b_{j-1} \text{ for } 1 \leq j \leq r_{k+1} - 1$$

$$a_{r_1+\ldots+r_k+r_{k+1}} < b_{r_{k+1}-1} \leq n$$

Similarly we divide $\underline{b}$ into initial segments of $i\underline{a}$ ($2 \leq i \leq n$) and denote their lengths by $s_1, s_2, \ldots$ ($s_i \geq 1$).

<u>Lemma 6.</u> There are maps $P, Q : \mathbb{N} \to \mathbb{N} \cup \{0\}$ such that

$$r_k = 1 + s_1 + \ldots + s_{P(k)} \text{ for } k \geq 1$$

$$s_k = 1 + r_1 + \ldots + r_{Q(k)} \text{ for } k \geq 1$$

<u>Proof.</u> We prove only the first statement. Suppose $1 + s_1 + \ldots + s_m < r_k < 1 + s_1 + \ldots + s_{m+1}$ for some m. We have

$a_{r_1+\ldots+r_{k-1}+j} = b_{j-1}$ for $1 \leq j \leq r_k - 1$ and $a_{r_1+\ldots+r_k} < b_{r_k-1}$

and

$b_{s_1+\ldots+s_m+l} = a_{l-1}$ for $1 \leq l \leq s_{m+1} - 1$.

Then

$a_{r_1+\ldots+r_{k-1}+s_1+\ldots+s_m+l+1} = a_{l-1}$ for $1 \leq l \leq r_k - s_1 - \ldots - s_m - 2$

and $a_{r_1+\ldots+r_k} < a_{r_k - s_1 - \ldots - s_m - 2}$.

If $r_k - s_1 - \ldots - s_m - 2 = 0$, we have $a_{r_1+\ldots+r_k} < a_0 = 1$, a contradiction. If $r_k - s_1 - \ldots - s_m - 2 \geq 1$, we have $\sigma^{r_1+\ldots+r_{k-1}+s_1+\ldots+s_m+2}\underline{a} < \underline{a}$, a contradiction to lemma 1. The lemma is proved.

It is easy to see that $G_1 = [\sigma\underline{a}, \underline{b}]$, $G_n = [\underline{a}, \sigma\underline{b}]$ and $G_i = [\underline{a}, \underline{b}]$ for $2 \leq i \leq n-1$. Furthermore $G_{ij} = G_j$, if $2 \leq i \leq n-1$, by the recursion formula for the $G_{x_0 \dots x_{m-1}}$, we have deduced in chapter I. In order to compute G_{1j} we have to consider two cases. If $r_1 = 1$, then $a_1 < b_0$ and $G_{1j} = \emptyset$ for $j < a_1$ and $j > b_0$, $G_{1a_1} = [\sigma^2 \underline{a}, \underline{b}]$, $G_{1j} = [\underline{a}, \underline{b}] = G_j$ for $a_1 < j < b_0$ and $G_{1b_0} = [\underline{a}, \sigma\underline{b}] = G_n$. If $r_1 > 1$ then $a_1 = b_0$ and $G_{1j} = \emptyset$ for $j \neq a_1$ and $G_{1a_1} = [\sigma^2 \underline{a}, \sigma\underline{b}]$. If $r_1 > 2$, we have also $a_2 = b_1$ and again only one successor $G_{1a_1a_2}$ and so on until $G_{1a_1 \dots a_{r_1-1}}$, because $a_i = b_{i-1}$ for $1 \leq i \leq r_1 - 1$, which has successors $G_{1a_1 \dots a_{r_1}} = [\sigma^{r_1+1} \underline{a}, \underline{b}]$, $G_{1a_1 \dots a_{r_1-1} j} = [\underline{a}, \underline{b}] = G_j$ for $a_{r_1} < j < b_{r_1-1}$ and $G_{1a_1 \dots a_{r_1-1} b_{r_1-1}} = [\underline{a}, \sigma^{r_1} \underline{b}] = G_{b_0 \dots b_{r_1-1}}$, because of lemma 6. Hence in any of the two cases we have got only one new interval, namely $G_{a_0 \dots a_{r_1}} = [\sigma^{r_1+1} \underline{a}, \underline{b}]$. Starting with this interval instead of G_1 above, we can do the same again and get that $G_{a_0 \dots a_j}$ $(j \geq r_1)$ has only one successor $G_{a_0 \dots a_{j+1}}$ as long as $j < r_1 + r_2 - 1$. As above $G_{a_0 \dots a_{r_1+r_2-1}} = [\sigma^{r_1+r_2} \underline{a}, \sigma^{r_2-1} \underline{b}]$ has successors $G_{a_0 \dots a_{r_1+r_2}} = [\sigma^{r_1+r_2+1} \underline{a}, \underline{b}]$, a new interval, G_j for $a_{r_1+r_2} < j < b_{r_2-1}$ and $G_{b_0 \dots b_{r_2-1}} = [\underline{a}, \sigma^{r_2} \underline{b}]$, again using lemma 6.

We can continue in this way and we can do the same for the $G_{b_0 \dots b_{j-1}}$. For convenience denote $(a_{i-1}, G_{a_0 \dots a_{i-1}})$ by c_i, $(b_{i-1}, G_{b_0 \dots b_{i-1}})$ by d_i $(i \geq 1)$ and (j, G_j) by e_j $(2 \leq j \leq n-1)$. Then $D = \{c_i, d_i : i \geq 1\} \cup \{e_2, \dots, e_{n-1}\}$ and M is given by the following arrows

$$c_i \to c_{i+1}, \ d_i \to d_{i+1} \quad (i \geq 1)$$

$$c_{r_1 + \dots + r_k} \to d_{r_k} \text{ and } e_j \text{ for } a_{r_1 + \dots + r_k} < j < b_{r_k - 1} \quad (k \geq 1)$$

$$d_{s_1 + \dots + s_k} \to c_{s_k} \text{ and } e_j \text{ for } a_{s_k - 1} < j < b_{s_1 + \dots + s_k} \quad (k \geq 1)$$

$$e_i \to c_1, \ d_1 \text{ and } e_j \text{ for } 2 \leq j \leq n-1 \quad (2 \leq i \leq n-1)$$

2. Now we return to maximal measures, which are Markov given by pairs of a left and a rigth eigenvector of M for the eigenvalue r(M).

For $g \in D$ let v_g be the length of the interval $\subset I$, which corresponds via φ to the interval represented by g. This interval is the disjoint union of all intervals, which are mapped by T to intervals corresponding to the elements of D following after g in M (cf. the definition of M). Hence $\beta v_g = \sum_{h \in D} M_{gh} v_h$. This means that $r(M) = \beta$ (i.e. $h_{top}(T) = \log \beta$) and v is a right eigenvector of M for the eigenvalue r(M). $\|v\|_\infty \le 1$.

We consider two cases. The first one is $n \ge 4$ and $n = 3$, if there is an arrow from $D \setminus \{e_2\}$ to $\{e_2\}$. If $n \ge 4$ it is easy to see that the irreducible submatrix of M, whose index set contains $\{e_2, \dots, e_{n-1}\}$, has a spectral radius strictly greater than the spectral radius of every other irreducible submatrix. If $n = 3$ and there is an arrow from $D \setminus \{e_2\}$ to $\{e_2\}$, say from c_k, then the irreducible submatrix of M, whose index set contains $\{e_2, c_1, \dots, c_k\}$ has spectral radius strictly greater than the spectral radius of every other irreducible submatrix. This is proved in [2]. Hence we have a unique irreducible submatrix M_1 with maximal spectral radius, i.e. we have unique maximal measure. It is Markov given by a right and left eigenvector of M_1. Take this left eigenvector (it can be extended to an eigenvector of M), which can be shown to be in l^1, and v above. This pair of vectors gives rise to a Markov measure with entropy log r(M), i.e. maximal entropy. Furthermore it assigns positive mass to every cylinder set. Hence the support of this measure is all of Σ_M. As this measure must be the unique maximal measure with support Σ_{M_1}, we have $M_1 = M$ and M is irreducible.

The second case is $n = 2$ and $n = 3$, if $\{e_2\}$ is an irreducible subset. Taking away $\{e_2\}$ from D means to take away those points from Σ_M, which have only e_2's left of some coordinate. Hence we have to consider the set $\{c_i, d_i : i \ge 1\}$, which we denote again by D. If M restricted to this set is reducible, we can decompose D into two disjoint subsets D_1 and D_2 such that there may be transition from D_1 to D_2, but not from D_2 to D_1. It is easy to see that there are $l \geq 1$ and $m \geq 1$ such that $D_1 = \{c_i, d_j : i < l,\ j < m\}$ and $D_2 = D \setminus D_1$. It is proved in [2] that $r(M|D_1) = 1$ or $r(M|D_1) > r(M|D_2)$. Choose m and l maximal such that $r(M|D_1) = 1$. The same arguments as above show that the maximal measure is again unique, that $M|D_2 = M_2$ is irreducible, and that the support of the maximal measure is Σ_{M_2}.

One can bring these results back to (I,T). We have

<u>Theorem 3.</u> T: $x \to \beta x + \alpha \pmod 1$ has topological entropy $\log \beta$ and unique maximal measure μ. It is absolutely continuous with respect to Lebesgue measure with density $h(x) = \sum_{g \in D} u_g 1_g$, u a left eigenvector of M and 1_g the characteristic function of the interval corresponding to $g \in D$.

<u>Proof.</u> We have to prove only the last assertion. Let $x_0 \ldots x_{m-1}$ be an admissible block in Σ_T^+. Then ${}_0[x_0 \ldots x_{m-1}] = \bigcup_{g \in D} H_g$, where $H_g = {}_0[x_0 \ldots x_{m-1}] \cap \{\underline{z} : \underline{y} = \psi'(\underline{z}) \text{ has } y_0 = g\}$. For every g with $H_g \neq \emptyset$ there are uniquely determined $g_1, \ldots, g_{m-2}, h \in D$ with $y_i = g_i$ $(1 \leq i \leq m-2)$ and $y_{m-1} = h$ for all $\underline{y} = \psi'(\underline{z})$, $\underline{z} \in H_g$. $\mu(H_g) = \pi_g P_{gg_1} \ldots P_{g_{m-2}h} = u_g v_h \beta^{-m+1}$. Hence $\mu({}_0[x_0 \ldots x_{m-1}]) = \sum_g u_g v_h \beta^{-m+1}$.

Because v_h is the length of the interval corresponding to h we have that $v_h \beta^{-m+1}$ is the length of the intersection of the interval corresponding to g and $\varphi^{-1}({}_0[x_0 \ldots x_{m-1}])$. Hence μ is absolutely continuous with respect to Lebesgue measure with density $\sum_{g \in D} u_g 1_g$.

In the first case above the support of the maximal measure is all of I, as we have seen. In the second case the support of the maximal measure is the union of the intervals corresponding to $c_l, c_{l+1}, \ldots, c_{l+m-1}$, $d_m, d_{m+1}, \ldots, d_{m+l-1}$. The intervals corrsponding to c_i and d_i for $i \geq m+1$ are contained in these intervals.

References

1. F. Hofbauer, β-shifts have unique maximal measure. Monatsh. f. Mathematik 85 (1978) 189 - 198.

2. F. Hofbauer, Maximal measures for simple piecewise monotonic transformations. Preprint.

3. W. Krieger, On the uniqueness of the equilibrium state. Math. Systems Theory 8 (1974) 97 - 104.

4. Y. Takahashi, Isomorphism of β-automorphisms to Markov automorphisms. Osaka Journal 10 (1973) 175 - 184.

Franz Hofbauer
Institut für Mathematik der Universität Wien
Strudlhofgasse 4
A-1090 Wien
AUSTRIA

A VARIATIONAL PRINCIPLE FOR THE TOPOLOGICAL CONDITIONAL ENTROPY

F.Ledrappier.

INTRODUCTION :

We consider continuous transformations of compact Hausdorff spaces into themselves. The topological conditional entropy is an invariant introduced by M.Misiurewicz, which is bigger than the defect of upper semi-continuity of the measure-theoretical entropy regarded as a function of invariant regular probability measures. Generally, it is strictly bigger (cf.(3) example 6.4.). Here we show that the topological conditional entropy can be defined by means of the entropy function on the cartesian square of the system. The formula we get looks like a variational principle for the t.c.e.. It implies that t.c.e. is conserved by principal extensions or by almost-conjugacy. It gives also a characterization of asymptotically h-expansive systems.

I. NOTATIONS AND RESULTS :

Let X be a compact Hausdorff space, T a continuous map from X into itself. The topological conditional entropy h^* is defined by M.Misiurewicz as follows :

Let A be a cover of X, Y a subset of X ; we denote N (Y,A) the smallest number of elements of A we need to cover Y.

Let A, B be covers of X ; we denote N (A/B) the biggest value of N (b,A) for b in B, h (A/B) the limit of the sequence

$\frac{1}{n}$ Log N ($\bigvee_{0}^{n-1} T^{-i} A / \bigvee_{0}^{n-1} T^{-i} B$), h (I/B) the supremum of h(A/B)

over all A open cover of X, and h^* is defined as the infimum of h (I/B) over all B open cover of X.

Let Y be another compact Hausdorff space, S a continuous map from Y into itself. We consider the product space XxY with the projection maps π_X and π_Y, and the product map T x S.

Let μ be a Radon probability measure on X x Y, we denote also μ its unique regular Borel extension to the σ-algebra $\mathcal{B}$ of Borel subsets of X x Y. Let $\mathcal{C}$ be the σ-algebra of Borel subsets of Y and $E_\mu^{\mathcal{C}}$ denote the conditional expectation with respect to π_Y^{-1} ($\mathcal{C}$).

Let η be the real function defined by η (t) = -t Logt, η (o) = o.

For a finite Borel partition ξ of X x Y, ξ = ($\xi_1 . \xi_2$ ...), we define :

$$H_\mu \ (\xi/Y) = \int \sum_i \ (E_\mu^{\mathcal{C}} 1_{\xi_i}) \ d\mu \ .$$

If the Radon probability μ is invariant by TxS, we denote $h_\mu(\xi/Y)$ the limit of the sequence $\frac{1}{n} H_\mu(\bigvee_{o}^{n-1} (SxT)^{-i} \xi/Y)$,

and $h(\mu/Y)$ the conditional metric entropy : $h(\mu/Y) = \sup_\xi h_\mu(\xi/Y)$.
We consider the set M(YxX, SxT) of all invariant Radon probability measures, with the weak topology, and we define $h^*(m/Y)$ by :

$$h^*(m/Y) = \limsup_{\mu \to m} h(\mu/Y) - h(m/Y) \text{ if } h(m/Y) \text{ is finite}, \infty \text{ otherwise.}$$

The main result is the following variational principle :

THEOREM 1 :

Let T be a continuous transformation of a compact Hausdorff space X. For any system (Y,S) the function $h^*(./Y)$ defined on M(YxX) is bounded by the topological conditional entropy h^*, and if the system (Y,S) is a copy of (X,T), then h^* is in fact a maximum for the function $h^*(./Y)$.

Theorem 1 follows from propositions 2 and 4. If h^* vanishes, we have :

THEOREM 2 :

Let T be a continuous transformation of a compact Hausdorff space X. The three following properties are equivalent :

i) $h^* = o$

ii) for any system (Y,S), the function h(./Y) is upper semi-continuous on M (YxX, SxT).

iii) let (X_1, T_1), (X_2, T_2) be two copies of (X,T). The function $h(./X_1)$ is upper semi-continuous on $M(X_1 x X_2, T_1 x T_2)$.

Such systems are called asymptotically h-expansive (cf (2),(3)).

From theorem 2 and Misiurewicz's example 6.4. in (3), there exists a system (X,T) for which the entropy is upper semi-continuous on M(X,T), and not on M(XxX, TxT). The property ii) in theorem 2 is called weakly expansive in (4).

The system (X,T) is said to be a factor of the system (Y,S) if there exists a continuous surjection π from Y onto X with $\pi S = T \pi$. We shall say (X,T) is a principal factor of (Y,S), or (Y,S) is a principal extension of (X,T) if for any measure in M(Y,S), h(m/X) is 0.

THEOREM 3 :

Topological conditional entropy is conserved by a principal extension.

Proof : If the entropy is not bounded on M(Y,S), neither is it on M(X,T) and for both systems, the t.c.e. is therefore infinite ((3) prop.3.3.). So we may suppose that the entropy is bounded on M(Y,S).

Let us consider (Y_1, S_1) and (Y_2, S_2) two copies of the system (Y,S), and X_i, T_i, π_i, i = 1,2 the corresponding objects. On the product space $Y_1 \times Y_2$ we call ϕ the product map $\pi_1 \times \pi_2$ onto $X_1 \times X_2$. Theorem 3 follows then readily from theorem 1, the continuity of the application $m \to m \circ \phi^{-1}$ from $M(Y_1 x Y_2, S_1 x S_2)$ onto $M(X_1 \times X_2, T_1 \times T_2)$, and the following remark :

If (Y,S) is a principal extension of (X,T), for any measure m in $M(Y_1 \times Y_2, S_1 \times S_2)$, we have : $h(m/Y_1) = h(m \circ \phi^{-1}/X_1)$.

Two systems are said principally equivalent if they have a common principal extension. Topological entropy and topological conditional entropy are invariants for that relation. If two systems are almost-conjugate ((1)), they have a common finite extension ;

therefore :

<u>Corrollary</u> : Topological conditional entropy is an invariant for almost-conjugacy.

A natural question here is whether two asymptotically h-expansive systems with the same topological entropy are principally equivalent.

II. <u>PROOFS</u> :

We first state some straightforward properties of the function $H_\mu(\xi/Y)$:

<u>Lemma 1</u> : Let $\mathcal{C}_\alpha$ be an increasing filter of finite sub σ-algebras of $\mathcal{C}$, which generate $\mathcal{C}$.

We have : $H_\mu(\xi/Y) = \inf_\alpha \int \sum_i (E_\mu^{\mathcal{C}_\alpha} 1_{\xi_i})\, d_\mu$.

<u>Lemma 2</u> : Let ξ be a finite partition, made of sets with a μ-negligible boundary. The function $\mu \to H_\mu(\xi/Y)$, defined on the set of Radon probability measures on YxX, is upper semi-continuous at μ.

<u>Lemma 3</u> : Let ξ and ξ' be two Borel partitions ; we have :

$H_\mu(\xi/Y) \leq H_\mu(\xi'/Y) + \operatorname{Log} N(\xi/\xi')$.

The two following propositions are conditional version of Misiurewicz's estimations ((3) proposition and theorem 4.2.) :

<u>Proposition 1</u> : Let ξ be a finite Borel partition of X, μ be in M(YxX, SxT), we have

$h(\mu/Y) \leq h_\mu(\pi_X^{-1}(\xi)/Y) + h(T/\xi)$.

Proof : Let ζ be a finite Borel partition of $X, \zeta = (\zeta_1, \ldots, \zeta_p)$.

We choose b_i, $i = 1, \ldots, p$, compact subset of ζ_i, such that :

(1) $h_\mu(\pi_X^{-1}(B)/Y) \geq h_\mu(\pi_X^{-1}(\xi)/Y) - 1$, where B is the partition made of $b_0 = X \setminus \bigcup_i b_i, b_1, \ldots, b_p$.

By lemma 3 we know that $H_\mu(\pi_X^{-1}(B)/Y)$ is not bigger than $H_\mu(\pi_X^{-1}(\xi)/Y) + \text{Log}\, N(B/\xi)$.

Applying this to partitions $\bigvee_0^{n-1} T^{-i} B$ and $\bigvee_0^{n-1} T^{-i} \xi$, dividing by n and letting n go to ∞, we get :

(2) $h_\mu(\pi_X^{-1}(B)/Y) \leq h_\mu(\pi_X^{-1}(\xi)/Y) + h(B/\xi)$.

We call C the open cover of X made of the sets $b_0 \cup b_i$, $i = 1, \ldots, p$.

We have $h(B/C) \leq \text{Log}\, 2$ by definition and $h(B/\xi) \leq h(C/\xi) + h(B/C)$.

Therefore :

(3) $h(B/\xi) \leq h(T/\xi) + \text{Log}\, 2$.

By (1), (2) and (3), we get, for any finite Borel partition ζ :

$h_\mu(\pi_X^{-1}(\zeta)/Y) \leq h_\mu(\pi_X^{-1}(\xi)/Y) + h(T/\xi) + \text{Log}\, 2 + 1$.

Hence we have :

$h(\mu/Y) \leq h_\mu(\pi_X^{-1}(\xi)/Y) + h(T/\xi) + \text{Log}\, 2 + 1$.

If we apply this formula to the transformation $S^m \times T^m$ and to the partition $\bigvee_0^{m-1} T^{-i} \xi$ for bigger and bigger m, we get the proposition.

Proposition 2 : For any measure m in $M(Y\times X, S\times T)$, $h^*(m/Y) \leq h^*$.

Proof :

Let A be a finite open cover of X and C be a finite Borel partition of X, made of sets with m-negligible boundary, and which as a cover refines A. We denote C^n the partition $\bigvee_{0}^{n-1} T^{-i} C$.

For any measure μ in $M(Y\times X, S\times T)$, we have by proposition 1 :

$$h(\mu/Y) \leq h_\mu(\pi_X^{-1}(C)/Y) + h(T/C) \leq \frac{1}{n} H_\mu(\pi_X^{-1}(C^n)/Y) + h(T/A).$$

Therefore for any n, we have :

$$\limsup_{\mu\to m} h(\mu/Y) \leq h(T/A) + \frac{1}{n}\limsup_{\mu\to m} H_\mu(\pi_X^{-1}(C^n)/Y)$$

$$\leq h(T/A) + \frac{1}{n} H_m(\pi_X^{-1}(C^n)/Y) \text{ by lemma 2,}$$

Therefore, we have

$$\limsup_{\mu\to m} h(\mu/Y) \leq h(T/A) + h(m/Y), \text{ for any open cover A of X.}$$

This shows proposition 2.

The converse part of theorem 1 follows from propositions 3 and 4 :

Proposition 3 : Let (X_1, T_1), (X_2, T_2) be two copies of (X,T) and $A = \{a_1, a_2, \ldots, a_p\}$ be an open cover of X. There exists a measure μ_A in $M(X_1\times X_2, T_1\times T_2)$ such that

i) $h(\mu_A/X_1) \geq h(T/A) - \frac{1}{p}$,

ii) μ_A is carried by the set $\bigcup_1^p \bar{a}_i^1 \times \bar{a}_i^2$.

<u>Proof</u> : Let us choose an open cover B of X such that

$h(B/A) \geq h(T/A) - \frac{1}{p}$, an atom a of the cover $A^n = \bigvee_0^{n-1} T^{-i}A$ such

that $N(a,B^n) = N(B^n/A^n)$, and x a point in a.

We choose now a neighbourhood L of the diagonal in $X \times X$ which is a Lebesgue number for B (that means that for any point y in X, there exists an open set b_y in B which contains all the points z with (y,z) in L).

A subset E of X is called (n,L)-separated if for any two different points x and x' in E, one of the pairs $(T^i x, T^i x')$, $i = 0, \ldots, n-1$, is not in L. We now choose E_n a maximal (n,L)-separated subset of a. By maximality the set

$$\bigcup_{y \in E_n} \left(\bigcap_0^{n-1} T^{-i} b_{T^i y} \right)$$ covers a and therefore the subset E_n has

at least $N(a,B^n)$ elements.

We consider the measures σ_n and μ_n on the space $X_1 \times X_2$ defined by :

$$\sigma_n = \frac{1}{\text{card } E_n} \sum_{y \in E_n} \delta_{(x,y)} \quad \text{and} \quad \mu_n = \frac{1}{n} \sum_{i=0}^{n-1} (T_1 \times T_2)^i \sigma_n .$$

We take μ_A being some measure adherent to the sequence μ_n.

The measure μ_A is invariant by $T_1 \times T_2$, we have to check that μ_A satisfies i) and ii).

Let ξ be a finite partition of X, $\xi = \{\xi_1, \ldots, \xi_q\}$ such that each ξ_i has a $\mu_A \circ \pi_{X_2}^{-1}$ negligible boundary and moreover each $\xi_i \times \xi_i$ is contained in L. We denote $\xi^p = \bigvee_0^{p-1} (T \times T)^{-i} \pi_{X_2}^{-1}(\xi)$.

As two different points of E_n cannot be in the same atom of

$\bigvee_{0}^{n-1} T^{-i} \xi$, we have :

$$H_{\sigma_n}(\xi^n/X_1) = \text{Log card } E_n.$$

For $0 \le j < m < n$, we cut the segment $(0, n)$ into the disjoint union of $(\frac{n}{m}) - 2$ segments $(j, j+m), \ldots (j+km, j+(k+1)m), \ldots$ and less than $3m$ other points. We have :

$$H_{\sigma_n}(\xi^n/X_1) \le \sum_{k=0}^{(\frac{n}{m})-2} H_{\sigma_n}\left(\bigvee_{i=j+km}^{j+(k+1)m-1} (T_1 \times T_2)^{-i} \pi_{X_2}^{-1}(\xi)/X_1\right) + 3m \text{ Log } q$$

$$\le \sum_{k=0}^{(\frac{n}{m})-2} H_{(T_1 \times T_2)^{j+km}\sigma_n}(\xi^m/X_1) + 3m \text{ Log } q,$$

and by summing over j, $0 \le j < m$:

$$m H_{\sigma_n}(\xi^n/X_1) \le \sum_{k=0}^{n-1} H_{(T_1 \times T_2)^k \sigma_n}(\xi^m/X_1) + 3m^2 \text{ Log } q$$

$$\le n H_{\mu_n}(\xi^m/X_1) + 3 m^2 \text{ Log } q \text{ by concavity of } \eta .$$

By comparing all the inequalities we have got, we have :

$$\frac{1}{m} H_{\mu_n}(\xi^m/X_1) \ge \frac{1}{n} \text{Log } N(B^n/A^n) - \frac{3m}{n} \text{ Log } q.$$

As the measure μ_A is adherent to the measures μ_n and by lemma 2,

$$\frac{1}{m} H_{\mu_A}(\xi^m/X_1) \geq \liminf_n \frac{1}{n} \operatorname{Log} N(B^n/A^n) = h(B/A).$$

By the choice of B, letting m go to infinity shows that the measure μ_A satisfies property i).

On the other hand, by definition of μ_n, we have :

$$\mu_n\left(\bigcup_i (a_i^1 \times a_i^2)\right) = \frac{1}{n}\sum_{j=0}^{n-1} \sigma_n\left(\bigcup_i (T_1 \times T_2)^{-j}(a_i^1 \times a_i^2)\right)$$

$$\geq \frac{1}{n}\sum_{j=0}^{n-1} \sigma_n(a^1 \times a^2) = \sigma_n(a^1 \times a^2) = 1.$$

That means that the support of every measure μ_n is contained in the closure of the set $\bigcup_i a_i^1 \times a_i^2$, which is $\bigcup_i \overline{a_i^1} \times \overline{a_i^2}$;

Therefore the measure μ_A satisfies also property ii).

<u>Proposition 4</u> : Let (X_1, T_1), (X_2, T_2) be two copies of (X,T). There exists a measure m in $M(X_1 \times X_2, T_1 \times T_2)$; carried by the diagonal, and such that : $h^*(m/X_1) = h^*$.

<u>Proof</u> :

For any finite open cover A of X, we choose the measure μ_A according to proposition 3, and we take m adherent to the measures μ_A when A becomes finer and finer.

By the property i) we have :

$$\limsup_{\mu \to m} h(\mu/X_1) \geq \liminf_{A} h(\mu_A/X_1) > \inf_{A} h(T/A) = h^*$$

On the other hand by property ii), the measure m is carried by the intersection of the sets $\cup_i \overline{a}_i^1 \times \overline{a}_i^2$, that is to say the diagonal of $X_1 \times X_2$. For every finite Borel partition ξ of X, the partitions $\pi_{X_1}^{-1}(\xi)$ and $\pi_{X_2}^{-1}(\xi)$ coïncide up to sets of m-measure 0, and then $h(m/X_1) = 0$.

Hence we have :

$$h^*(m/X_1) = \limsup_{\mu \to m} h(\mu/X_1) - h(m/X_1) \geq h^* ;$$

By proposition 2, $h^*(m/X_1) \leq h^*$. The measure m is the measure we wanted.

BIBLIOGRAPHIE

1. R.L.ADLER and B.MARCUS. Topological entropy and equivalence of dynamical systems. Preprint, to appear in: Memoirs Amer.Math.Soc.

2. R.BOWEN. Entropy-expansive maps. T.A.M.S. 164 (1972) 323-333.

3. M.MISIUREWICZ. Topological conditional entropy. Studia Math. LV (1976), 175-200.

4. F.LEDRAPPIER. Mesures d'équilibre d'entropie complètement positive. Astérisque 50 (1977), 251-272.

F. Ledrappier
Université de Paris VI
Lab. de Calcul des
Probabilités
4, place Jussien - Tour 56
F-75230 Paris cédex 05

WEAK MIXING FOR SEMI-GROUPS OF MARKOV OPERATORS WITHOUT FINITE INVARIANT MEASURES

by

Michael Lin

Department of Mathematics

Ben Gurion University of the Negev

Beer-Sheva, Israel

1. Introduction

There have been various attempts to generalize the notion of a weak mixing transformation to ergodic transformations without finite invariant measures. In (1) we obtained the following.

Weak mixing theorem: Let P *be a Markov operator on* $L_\infty(X,\Sigma,m)$. *Then the following are equivalent:*

(i) P *is ergodic and has no unimodular eigenvalues* $\neq 1$.

(ii) For every $u \in L_1$ *with* $\int u\,dm = 0$ *and every* $f \in L_\infty$ *we have*

$$\lim_{N\to\infty} \frac{1}{N} \sum_{n=1}^{N} |\langle u, P^n f\rangle| = 0.$$

(iii) For every ergodic Markov operator Q *with a finite invariant measure,* $P \times Q$ *is ergodic.*

It is the purpose of this note to indicate how to obtain the weak mixing theorem for semi-groups.

2. The weak mixing theorem for semigroups

Definition. A complex number re^{is} is an *eigenvalue* of a semi-group $\{T_t\}_{t\geq 0}$ of linear operators in a Banach space, if there exists a vector f such that $T_t f = r^t e^{ist} f$ for every t.

Theorem. *Let* $\{P_t\}_{t>0}$ *be a weak -* continuous semi-group of Markov operators on* $L_\infty(X,\Sigma,m)$. *Then the following are equivalent:*

(i) $\{P_t\}$ *is ergodic, and has no unimodular eigenvalues* $\neq 1$.

(ii) *For every* $u\in L_1$ *with* $\int u\,dm = 0$ *and every* $f\in L_\infty$, *we have*

$$\lim_{T\to\infty} \frac{1}{T} \int_0^T |\langle u, P_t f\rangle|\,dt = 0.$$

(iii) *For every ergodic weak-* continuous semi-group of Markov operators* $\{Q_t\}$ *with a finite invariant measure,* $\{P_t \times Q_t\}$ *is ergodic.*

The proof of *(iii)* ⇒ *(ii)* is a direct modification of the proof of theorem 4.4 in (1), and *(ii)* ⇒ *(i)* is trivial.

The steps in proving *(i)* ⇒ *(iii)* are similar to those of (1), but involve deeper constructions. The first step is the reduction to point transformations (semi-flows), which uses the shift for continuous-time Markov processes, and requires the analogue of theorem 2.7 of (1). The second step is a technical one: reduction to the separable case, and this is achieved as in (1, lemma 4.5), using the continuity of the L_1 semi-groups (and the separability of $\{t > 0\}$). The final step needs the following proposition (instead of (1, prop. 4.6)).

<u>Proposition</u>. *Let* $\{\theta_t\}$ *be a non-singular semi-flow on* (X,Σ,m). *Let* $\{U_t\}$ *be a continuous unitary group on a separable Hilbert space* H. *Let* $F(x)$ *be a measurable function from* X *into* H *such that for each* $t>0$, $F(\theta_t x) = U_t F(x)$ *a.e.. If* $\{\theta_t\}$ *has no unimodular eigenvalues* $\neq 1$, *then, for each* $t > 0$, $U_t F(x) = F(x)$ *a.e.*

<u>Proof</u>. We may assume that $||F(x)|| \leq k$ (see (1)). We let

$$H_o = \{h\varepsilon H : \lim_{\alpha\to\infty} \frac{1}{\alpha} \int_0^\alpha U_t h dt = 0\}.$$

Fix $h\varepsilon H_o$, and let $H_1 = \text{clm}\{U_t h: -\infty<t<\infty\}$. By Stone's theorem (2, XII. 6.1) and by the spectral representation for unbounded self-adjoint operators (2, §XII. 3), there is a positive measure η on $R = (-\infty,\infty)$ such that $H_1 \cong L_2(R,\eta)$, $(U_t h,h) = \int e^{its}\eta(ds)$ for $t\varepsilon R$, and U_t on H_1 corresponds to multiplication by the function $g_t(s) = e^{its}$.

Let E be the orthogonal projection on H_1, and define $F_1(x) = EF(x)$. Since H_1 is invariant under all $\{U_t\}_{t\varepsilon R}$, we have $U_t F_1(x) = F_1(x)$ a.e., for each $-\infty<t<\infty$.

By (2, III. 11.17) there is a bi-measurable $f(x,s)$ on $X\times R$ such that for a.e. x, $f(x,\cdot)$ represents $F_1(x)$ in $L_2(R,\eta)$.

Fix $t\varepsilon R$. Then for x in a set of full measure, $e^{its}f(x,s) = f(\theta_t x,s)$ for a.e.s. Hence for $m\times\eta$ a.e. (x,s), $e^{its}f(x,s) = f(\theta_t x,s)$. Take a sequence $\{t_j\}$ dense in R. Then $f(\theta_{t_j} x,s) = e^{it_j s}f(x,s)$ for a.e. (x,s) and each j. Let $f_s(x) = f(x,s)$. Then for a.e.s, $f_s o\theta_{t_j} = e^{it_j s}f_s$ for each j.

Fix s, and let $A = \{x \in X : |f_s(x)| \leq \alpha\}$. Then A is invariant for each θ_{t_j}, and weak-* continuity in L_∞ of $\{\theta_t\}$ shows that A is invariant for the semi-flow $\{\theta_t\}$. Hence $g_\alpha = 1_A f_s$ is in L_∞, and, by weak-* continuity of $\{\theta_t\}$, $g_\alpha o \theta_t = e^{ist} g_\alpha$. By assumption, for $s \neq 0$, $g_\alpha = 0$ in L_∞, so $f_s = 0$ a.e., for $s \neq 0$. However, since $h \in H_o$, $\eta\{0\} = 0$. Hence $f(s,x) = 0$ $m \times \eta$ a.e., so for a.e. x, $F_1(x) = 0$, or $F(x)$ H_1 a.e. By separability of H, we take a sequence $\{h_n\}$ dense in H_o to obtain that $F(x) \perp H_o$ for a.e. x, which shows that F(x) is invariant for the unitary group $\{U_t\}$.

The proof of the theorem, based on the proposition, is similar to (1).

References

1. J. Aaronson, M. Lin and B. Weiss: Mixing properties of Markov operators and ergodic transformations, and ergodicity of Cartesian products (to appear).
2. N. Dunford and J.T. Schwartz: Linear operators, Interscience, New-York; part I, 1958; part II, 1963.

ERGODIC GROUP AUTOMORPHISMS AND SPECIFICATION

D. A. Lind
University of Washington
Seattle, Washington 98195

§1. Introduction.

I first want to discuss how I became interested in finding out which ergodic automorphisms of compact groups satisfy a property called specification, and then describe the answer for ergodic toral automorphisms. This settles a question raised by Sigmund [15, §2]. One consequence of the answer is that Markov partitions, such as those found by Adler and Weiss [1] for two-dimensional toral automorphisms and by Bowen [3] for hyperbolic toral automorphisms, cannot be constructed for nonhyperbolic automorphisms. There seem to be essential differences between the dynamic behavior of hyperbolic and nonhyperbolic automorphisms. Some questions that can be answered in the hyperbolic case using the Markov partition machinery, e.g. concerning the distribution of periodic orbits, remain open for nonhyperbolic toral automorphisms.

§2. Splitting skew products.

When dealing with the ergodic properties of group automorphisms, transformations of the following form often arise. Let $U:X \to X$ be an invertible measure-preserving transformation (hereafter shortened to "map") of a Lebesque space (X,μ), $S:G \to G$ be a (continuous, algebraic) automorphism of a compact metrizable group G written additively, and $\alpha:X \to G$ be a measurable function. Form the skew product $U \times_\alpha S:X \times G \to X \times G$ defined by $(U \times_\alpha S)(x, g) = (Ux, Sg + \alpha(x))$. Such skew products arise, for example, when there is a closed subgroup H of G that is invariant under S. For by taking a Borel cross-section to the projection $G \to G/H$, the automorphism S can be written as the skew product of the factor automorphism $S_{G/H}$ with the restriction S_H of S to H (see [8, p. 209] for details).

While investigating the Bernoullicity of group automorphisms in [8], I noticed that by using Thouvenot's relative isomorphism theory one could show that skew products with ergodic group automorphisms are always isomorphic to direct products, via an isomorphism that preserves the group fibers. This means that there is a map $W:X \times G \to X \times G$ of the form $W(x,g) = (x, W_x(g))$ such that

(1) $$(U \times_\alpha S)W = W(U \times S).$$

Unfortunately, Thouvenot's theory gives no information about the individual maps $W_x:G \to G$. Demanding that they be group translations, i.e. $W_x(g) = g + \beta(x)$, when substituted into (1), amounts to solving the functional equation

(2) $$\alpha(x) = \beta(Ux) - S\beta(x),$$

where α, U, and S are known, and the measurable function $\beta:X \to G$ is to be found. It turns out that this is always possible.

<u>Splitting Theorem [9]</u>. If S is an ergodic group automorphism, then for arbitrary α and U the functional equation (2) can be solved for β.

One application of this result is the simplest proof so far of Katznelson's result [7] that ergodic toral automorphisms are Bernoulli (see [9, §5]), one that avoids the Diophantine approximation arguments used in previous proofs. Other applications are mentioned in [9, §1].

Bowen's property "specification" plays a key role in the proof of the Splitting Theorem given in [9], so let me first review this property, and then indicate its use.

Let (Y,d) be a compact metric space, and $f:Y \to Y$ be continuous. The transformation f <u>obeys specification</u> if for every $\varepsilon > 0$ there is an $M(\varepsilon)$ such that for every $r \geq 2$ and r points $y_1, \cdots, y_r \in Y$, and for every set of integers $a_1 \le b_1 < a_2 \le b_2 < \cdots < a_r \le b_r$ and p with $a_i - b_{i-1} \ge M(\varepsilon)$ $(2 \le i \le r)$ and $p \ge b_r - a_1 + M(\varepsilon)$, there is a point $y \in Y$ with $d(f^n y, f^n y_i) < \varepsilon$ for $a_i \le n \le b_i$, $1 \le i \le r$, and with $f^p(y) = y$. Basically this definition means that given specified pieces $\{f^n y_i : a_i \le n \le b_i\}$ of orbits of different points for disjoint blocks of times, if there is enough space between these blocks then these pieces can be well approximated during the same time blocks by the orbit of a single periodic point.

Several important dynamical systems obey specification, including hyperbolic toral automorphisms and subshifts of finite type (see [5, Ch. 21]). Bowen introduced this property to produce the unique equilibrium measure for an Axiom A diffeomorphism [3, Ch. 4] and to construct Markov partitions for such diffeomorphisms [3, Ch. 3]. Sigmund [12], [13], [14], [15] and Kamae [6] have used variants of specification to study orbits and generic properties of invariant measures to generalize number-theoretic facts about decimal expansions.

Actually, solving (2) involves only the orbit copying part of specification. Hence say that $f:Y \to Y$ <u>obeys weak specification</u> if it satisfies the definition of specification except for the periodic point condition $f^p(y) = y$. Weak specification was used by Ruelle [11] in studying the statistical mechanics of lattice actions.

Let me now sketch how to solve the functional equation (2) for those automorphisms S obeying weak specification. Let S be an ergodic automorphism of a compact abelian group G, and equip G with a translation invariant metric ρ. Suppose that S satisfies weak specification on (G, ρ). Let $U:X \to X$ and $\alpha:X \to G$ be measurable. I will find β by constructing approximating solutions β_k defined on successively larger parts of X.

To begin, choose positive ε_k with $\Sigma\, \varepsilon_k < \infty$. Let $M(\varepsilon_1)$ be given by the weak specification property of S, and choose an integer h_1 so that $M(\varepsilon_1)/h_1 < \varepsilon_1$. Let $F_1 \subset X$ be a Rohlin base for U of height $h_1 + M(\varepsilon_1)$, and let $E_1 = \cup\, \{U^j F_1 : 0 \le j < h_1\}$. The base F_1 can be chosen so that $\mu(E_1) > 1 - 3\, \varepsilon_1$. Define β_1 arbitrarily but measurably on F_1. The functional equation (2) forces the definition of β_1 inductively up the stack. Specifically, for

$x \in F_1$,

$$\beta_1(Ux) = S\beta_1(x) + \alpha(x) ,$$
$$\beta_1(U^2x) = S\beta_1(Ux) + \alpha(Ux)$$
$$= S^2\beta_1(x) + S\alpha(x) + \alpha(Ux),$$

and in general

$$\beta_1(U^jx) = S^j\beta_1(x) + \alpha_j(x) \quad (x \in F_1,\ 0 \le j < h_1)$$

where

$$\alpha_j(x) = \sum_{m=0}^{j-1} S^{j-m-1}\alpha(U^m x).$$

This defines β_1 on the stack E_1, and it satisfies (2) on all but the top level.

Similarly, for the given $\varepsilon_2 > 0$, find $M(\varepsilon_2)$, h_2, F_2, E_2, where h_2 is much larger than h_1. Now comes the essential point. Once β_2 is defined at $x \in F_2$, its definition on the orbit piece $\{U^jx\colon 0 \le j < h_2\}$ is forced by (2). However, β_1 is already defined on certain subpieces of this piece, namely the blocks of time when the orbit of x is in E_1. It may not be possible to select a value for $\beta_2(x)$ so that on the subpieces of the orbit where β_1 is already defined, β_2 will agree with, or even be close to, to previous function β_1. The role of specification is to show that such a selection is possible.

Suppose I let $\beta_2'(x) = g_0$, and see what the trouble is. Now β_1 is already defined on certain subpieces, say $\{U^nx\colon a_i \le n \le b_i, 1 \le i \le r\}$, where $b_i = a_i + h_1$, and by construction $a_{i+1} - b_i \ge M(\varepsilon_1)$. The definition of β_2' on

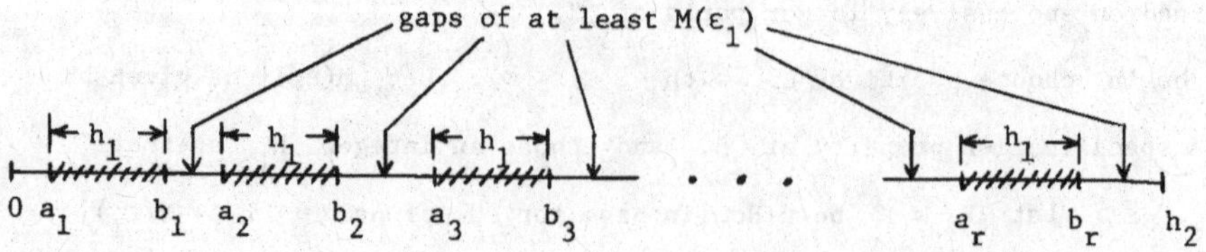

those pieces is forced:

$$\beta_2'(U^{a_i+j}x) = S^{a_i+j}g_0 + \alpha_{a_i+j}(x)$$

for $1 \le i \le r$, $0 \le j < h_1$. Also,

$$\beta_1(U^{a_i+j}x) = S^j\beta_1(U^{a_i}x) + \alpha_j(U^{a_i}x).$$

Since

$$\alpha_{a_i+j}(x) - \alpha_j(U^{a_i}x) = S^j\alpha_{a_i}(x),$$

subtraction gives

$$\beta_2'(U^{a_i+j}x) - \beta_1(U^{a_i+j}x) = S^j[S^{a_i}g_0 - \beta_1(U^{a_i}x) + \alpha_{a_i}(x)].$$

Since the bracketed expression is independent of j, the error on the subpiece $\{U^n x : a_i \le n \le b_i\}$ is the orbit of the point in brackets, different points for different i. The time gaps between these subpieces are at least $M(\varepsilon)$, so weak specification can be used to adjust our original choice of $\beta_2'(x)$ to decrease these errors to be uniformly less than ε_1. For by weak specification, there is a $g_1 \in G$ such that

$$\rho(S^{a_i+j}g_1, S^j[S^{a_i}g_0 - \beta_1(U^{a_i}x) + \alpha_{a_i}(x)]) < \varepsilon_1$$

for $0 \le j < h_1$, $1 \le i \le r$. Define $\beta_2(x) = \beta_2'(x) - g_1$. Then it is easy to check using translation invariance of ρ that if β_2 is defined up the stack using (2), then $\rho(\beta_2(U^n x), \beta_1(U^n x)) < \varepsilon_1$ whenever $U^n x \in E_1$, $0 \le n < h_2$.

Thus β_2 is defined on more of X, solves (2) where defined, and is uniformly close to β_1 where the latter is defined.

Similarly, construct β_k defined on E_k with $\mu(E_k) \to 1$, such that β_k solves (2) where defined, and such that β_{k+1} and β_k are within ε_k on E_k. Since $\sum \varepsilon_k < \infty$, $\{\beta_k\}$ is almost uniformly Cauchy, so converges to a measurable function β defined almost everywhere on X which solves (2). Details of this argument are in [9, §4].

§3. Specification.

If every ergodic group automorphism satisfied weak specification, then the simple argument in §2 would be all that is needed to solve (2). A general ergodic automorphism can be built up from certain basic automorphisms that do satisfy weak specification by using factors, products, inverse limits, and extensions by basic automorphisms (see [9, §7]). All but the last process preserve weak specification. Unfortunately, there are extensions of homeomorphisms obeying weak specification by a basic automorphism that do not obey weak specification. In fact, a toral automorphism S' is given below having an invariant subgroup H such that both $S'_{G/H}$ and S'_H obey weak specification, but S' does not.

Hence at least some of the complication in [9] of solving (2) for a general ergodic group automorphism seems intrinsic to the specification approach. Dan Rudolph [10] has found another way of solving (2) by using a relativised isomorphism theorem for measure-preserving actions of a skew product of the integers with a compact group.

Which group automorphisms obey specification?

The shift whose state space is a finite group certainly does, and I believe so do all ergodic automorphisms of totally disconnected compact abelian groups. This is true in many cases, but I have not yet found a general proof.

However, for ergodic toral automorphisms there is a complete answer. Such automorphisms come in three flavors, depending on the spectral properties of the associated linear map.

(1) Hyperbolic automorphisms have no eigenvalues on the unit circle (i.e. no unitary eigenvalues).

(2) Central spin automorphisms have some unitary eigenvalues, and on the eigenspace of the unitary eigenvalues the associated linear map is an isometry (with an appropriate metric); this means that the Jordan blocks for the unitary eigenvalues have no off-diagonal 1's.

(3) Central skew automorphisms are what's left, namely those with off-diagonal 1's in the Jordan block of some unitary eigenvalue.

Central spin and central skew automorphisms occur. Let

$$S = \begin{pmatrix} 0 & 1 & 0 & 0 \\ 0 & 0 & 1 & 0 \\ 0 & 0 & 0 & 1 \\ -1 & -4 & 2 & -4 \end{pmatrix}.$$

The eigenvalues for S are $\sqrt{2} - 1 \pm i\sqrt{2\sqrt{2} - 2}$ (both unitary), and $-\sqrt{2} - 1 \pm \sqrt{2\sqrt{2} + 2}$ ($\approx -4.61, -0.21$). Hence S is a central spin automorphism. Dimension 4 is least possible for such an automorphism. Also, if I denotes the 4×4 identity matrix, then

$$S' = \begin{pmatrix} S & I \\ 0 & S \end{pmatrix}$$

is a central skew automorphism. Dimension 8 is least possible for such an automorphism.

The specification behavior of each of these classes is different.

Theorem. (i) Hyperbolic automorphisms obey specification.

(ii) Central spin automorphisms obey weak specification, but never obey specification.

(iii) Central skew automorphisms never obey even weak specification.

Part (i) is due to Bowen [2]. The first part of (ii) is proved in [9] as part of the general solution to (2). In the next two sections I briefly indicate the geometric ideas behind (iii) and the negative part of (ii).

§4. Central skew automorphisms.

Let S be a central skew automorphism of the d-dimensional torus $\mathbb{T}^d = \mathbb{R}^d/\mathbb{Z}^d$, and denote its lifting to a linear map of $\mathbb{R}^d$ by the same symbol. Denote by $E^s, E^u, E^c \subset \mathbb{R}^d$ the stable, unstable, and central subspaces of S corresponding to the eigenvalues of S inside, outside, and on the unit circle, respectively. Then $\mathbb{R}^d = E^s \oplus E^c \oplus E^u$. It is not dangerous to identify these subspaces with their projections to $\mathbb{T}^d$ since by ergodicity of S they are disjoint from $\mathbb{Z}^d$.

Since S is central skew, there are disjoint 2-dimensional subspaces E_1^c and E_2^c of E^c such that the restriction of S to $E_1^c \oplus E_2^c$ has the form

$$\begin{pmatrix} Q & I \\ 0 & Q \end{pmatrix},$$

where Q is a rotation and I is the 2 × 2 identity matrix. For example, in the central skew automorphism S' given in §3,

$$\begin{pmatrix} \sqrt{2}-1 & \sqrt{2\sqrt{2}-2} \\ -\sqrt{2\sqrt{2}-2} & \sqrt{2}-1 \end{pmatrix}$$

with respect to an appropriate basis. E^c could be larger than $E_1^c \oplus E_2^c$. Note that the matrix of S^j on $E_1^c \oplus E_2^c$ is

$$(3) \qquad \begin{pmatrix} Q^j & jQ^{j-1} \\ 0 & Q^j \end{pmatrix}$$

For convenience, give $\mathbb{T}^d$ a translation-invariant product metric ρ inherited from the eigenspaces, and let $B^s(\varepsilon)$ denote the ε-ball around 0 in the stable subspace (projected to $\mathbb{T}^d$), and similarly define $B^u(\varepsilon)$, $B^c(\varepsilon)$, $B(\varepsilon)$ (for details, see [9, §6]).

Note that $u \in \mathbb{T}^d$ stays within ε of $t \in \mathbb{T}^d$ under the first n iterates of S precisely when u-t stays within ε of 0. Hence it is important to analyse

$$\begin{aligned} A(n,\varepsilon) &= \{t \in \mathbb{T}^d : \rho(S^j t, 0) < \varepsilon \text{ for } 0 \le j \le n\} \\ &= \bigcap_{j=0}^{n} S^{-j}B(\varepsilon) \\ &= A^s(\varepsilon,n) \oplus A^c(\varepsilon,n) \oplus A^u(\varepsilon,n), \end{aligned}$$

where $A^s(\varepsilon,n) = \cap_{j=0}^n S^{-j} B^s(\varepsilon)$, etc.

Since S^{-1} expands on E^s, $A^s(\varepsilon,n)$ is "essentially" $B^s(\varepsilon)$, i.e. There are absolute constants K_1 and K_2 independent of ε and n such that $K_1B^s(\varepsilon) \subset$

$A^s(\varepsilon,n) \subset K_2 B^s(\varepsilon)$.

Similarly, $A^u(\varepsilon,n)$ is "essentially" $S^{-n}B^u(\varepsilon)$.

For the central direction, let $t_i \in E_i^c$, so $t_1 \oplus t_2 \in E_1^c \oplus E_2^c$. Then from (3),

$$S^j(t_1 \oplus t_2) = (Q^i t_1 + jQ^{j-1}t_2) \oplus Q^j t_2 .$$

Since this is to be within ε of 0 for $0 \le j \le n$, and Q is an isometry, clearly $\rho(t_1, 0) < \varepsilon$, $\rho(t_2,0) < \varepsilon$, and $\rho(Q^j t_1 + jQ^{j-1}t_2, 0) < \varepsilon$. The last inequality forces $\rho(t_2, 0) < 2\varepsilon/j$, so in particular $\rho(t_2, 0) < 2\varepsilon/n$. This thinness is one eigendirection, decreasing with n, is the key geometrical fact in the proof.

These observations are enough to show that S does not obey weak specification. Fix an $\varepsilon > 0$ small enough so that the projection of $B(5\varepsilon) \subset \mathbb{R}^d$ to $\mathbb{T}^d$ is injective. Let $M > 0$ be given. I will find $t_1, t_2 \in \mathbb{T}^d$ and integers $0 = a_1 < b_1 < a_2 < b_2$ with $a_2 - b_1 = M$ such that no $t \in \mathbb{T}^d$ has $\rho(S^j t, S^j t_i) < \varepsilon$ for $a_i \le j \le b_2$, $i = 1, 2$.

The intersection $S^M B^u(\varepsilon) \cap [B^S(5\varepsilon) \oplus B^c(5\varepsilon)]$ is finite, say $\{s_1, \cdots, s_r\}$. For n sufficiently large, the projection P of $\cup_{i=1}^r \{B_2^c(\varepsilon/n) + s_i\}$ to $B_2^c(5\varepsilon)$ cannot cover all of $B_2^c(5\varepsilon)$. For such an n, choose an element $u \in B_2^c(5\varepsilon)$ and an integer m such that $u + B_2^c(\varepsilon/m)$ is disjoint from P.

Now let $t_1 = 0$, $t_2 = u$, $a_1 = 0$, $b_1 = n$, $a_2 = n + M$, and $b_2 = a_2 + m$. Suppose there were a $t \in \mathbb{T}^d$ with $\rho(S^j t, S^j t_i) < \varepsilon$ for $a_i \le j \le b_i$. Then $t \in A(\varepsilon,n)$ and $S^{n+M} t \in u + A(\varepsilon,m)$. Hence

$$S^{n+M} t \in [S^{n+M} A(\varepsilon,n)] \cap [u + A(\varepsilon,m)] .$$

Since the component of $A(\varepsilon,n)$ in the E_2^c direction is $B_2^c(\varepsilon/n)$, the projection to $B_2^c(5\varepsilon)$ of the first term in this intersection is contained in P, while that of the second is by construction disjoint from P. This contradiction completes the argument.

§5. Central spin automorphisms.

Central spin automorphisms obey weak specification [9, §6], but the argument here shows that they never obey specification. In fact, if S is such an automorphism of $\mathbb{T}^d$, then all sufficiently small $\varepsilon > 0$ will have the property that for every $M > 0$ there are $t_1 \in \mathbb{T}^d$ and $n > 0$ such that no $t \in \mathbb{T}^d$ has $\rho(S^j t, S^j t_1) < \varepsilon$ for $-n \leq t \leq 0$ and $S^{n+M} t = t$. This will contradict the specification definition for $r = 1$, $a_1 = -n$, $b_1 = 0$, and $p = n + M$.

Since S has a central spin factor with irreducible characteristic polynomial, and specification is preserved under factors, assume that S has irreducible characteristic polynomial. The purpose of this is to guarentee that $(E^s \oplus E^u) \cap \mathbb{Z}^d = \{0\}$, which follows from irreducibility because then S cannot preserve a nontrivial lattice.

Choose $\varepsilon > 0$ as small as in §4. Let $t_1 \in E^u$ such that $\rho(t_1, 0) = 2\varepsilon$.

Since S is central spin, its restriction to the central subspace E^c is an isometry, say Q. Hence the identity on E^c can be arbitrarily well approximated by arbitrarily large powers of Q.

Now $\{S^M[B^u(\varepsilon) + t_1]\} \cap \{B^s(5\varepsilon) \oplus B^c(5\varepsilon)\}$ is finite, say $\{s_1, \cdots, s_r\}$. Since $0 \notin S^M[B^u(\varepsilon) + t_1]$, and $(E^s \oplus E^u) \cap \mathbb{Z}^d = \{0\}$, the projection of $s_1, \cdots, s_r$ to $B^c(5\varepsilon)$ are all displaced at least some quantity $\delta > 0$ from 0. Choose n so that Q^{n+M} is so close to the identity that for every $s \in B^c(5\varepsilon)$ with $\rho(s,0) \geq \delta$, the map $u \mapsto s + Q^{n+M} u$ has no fixed points $u \in B^c(5\varepsilon)$.

Suppose there were a $t \in \mathbb{T}^d$ with $\rho(S^j t, S^j t_1) < \varepsilon$ for $-n \leq j \leq 0$ and $S^{n+M} t = t$. If u denotes the projection of $S^{-n} t$ to $B^c(5\varepsilon)$, then the projection of $S^M t = S^{n+M}(S^{-n} t)$ to $B^c(5\varepsilon)$ has the form $s_i + Q^{n+M} u$ for some i. Since $S^{-n} t = S^M t$, their projections u and $s_i + Q^{n+M} u$ must agree. But $\rho(s_i, 0) \geq \delta$, so there are no fixed points of the map $u \mapsto s_i + Q^{n+M} u$ for $u \in B^c(5\varepsilon)$, showing that such a t does not exist.

§6. Remarks.

Nonhyperbolic toral automorphisms seem to behave differently from the hyperbolic ones. For example, a modification of the geometric ideas here shows that for nonhyperbolic automorphisms, every fine enough partition is not weak Bernoulli, although every partition is very weak Bernoulli since the automorphism is a Bernoulli shift. This should be contrasted with Bowen's result [4] that for hyperbolic automorphisms every smooth partition is weak Bernoulli. The geometry also shows clearly certain limits to independence that forced Katznelson [7] to introduce the intermediate idea of "almost weak Bernoulli" in the first proof that ergodic toral automorphisms are Bernoulli. Details concerning these remarks will appear elsewhere.

It follows from the theorem in §3 that Markov partitions in the sense of Bowen [3] do not exist for nonhyperbolic toral automorphisms. For the existence of a Markov partition would imply that the automorphism is a factor of a Markov shift. Such shifts obey specification, and specification is trivially preserved under factors.

Thus nonhyperbolic toral automorphisms are examples of smooth systems for which the usual machinery of Markov partitions is unavailable, but which still can be analysed in detail. Yet many questions about them, which can be answered in the hyperbolic case, remain unsettled. Sample: Are the periodic orbit measures weakly dense in the space of invariant measures? In particular, is there a sequence of periodic orbits that converges weakly to Lebesgue measure, i.e. is uniformly distributed? Nobody seems to know.

References

1. R. Adler and B. Weiss, Similarity of automorphisms of the torus, Mem. Amer. Math. Soc., 98 (1970).

2. Rufus Bowen, Periodic points and measures for Axiom A diffeomorphisms, Trans. Amer. Math. Soc., 154 (1971), 377-397.

3. __________, Equilibrium States and the Ergodic Theory of Anosov Diffeomorphisms, Springer Lecture Notes in Math. 470, Berlin, 1975.

4. __________, Smooth partitions of Anosov diffeomorphisms are weak Bernoulli, Israel J. Math. 21 (1975), 95-100.

5. Manfred Denker, Christian Grillenberger, and Karl Sigmund, Ergodic Theory on Compact Spaces, Springer Lecture Notes in Math. 527, Berlin, 1976.

6. Teturo Kamae, Normal numbers and ergodic theory, Proc. 3rd Japan-USSR Symp. Prob. Th., Springer Lecture Notes in Math. 550 (1976), 253-269.

7. Y. Katznelson, Ergodic automorphisms of T^n are Bernoulli shifts, Israel J. Math. 10 (1971), 186-195.

8. D. A. Lind, The structure of skew products with ergodic group automorphisms, Israel J. Math. 28 (1977), 205-248.

9. __________, Split skew products, a related functional equation, and specification, to appear, Israel J. Math.

10. Daniel J. Rudolph, An isomorphism theory for Bernoulli free Z-skew-compact group actions, to appear.

11. David Ruelle, Statistical mechanics on a compact set with $\mathbb{Z}^2$ action satisfying expansiveness and specification, Trans. Amer. Math. Soc. 185, (1973), 237-251.

12. Karl Sigmund, Generic properties of invariant measures for Axiom A diffeomorphisms, Inventiones Math. 11 (1970), 99-109.

13. __________, Ergodic averages for Axiom A diffeomorphisms, Z. Wahrscheinlichkeitsth. verw. Geb. 20 (1971), 319-324.

14. __________, Mixing measures for Axiom A diffeomorphisms, Proc. Amer. Math. Soc. 36 (1972), 497-504.

15. __________, On dynamical systems with the specification property, Trans. Amer. Math. Soc. 190 (1974), 285-299.

Measures of Maximal Entropy for a Class of Skew Products

Brian Marcus and Sheldon Newhouse

University of North Carolina - Chapel Hill

1. Introduction.

Let U be a Borel measurable isomorphism of a compact metric space. We define the topological entropy $(h(U))$ of U to be

$$h(U) = \sup \{h_\mu(U): \mu \text{ is a } U\text{-invariant Borel probability measure}\}.$$

The well-known variational principle ([G], [DGS]) asserts that this definition agrees with the usual definition of topological entropy in the case that U is a homeomorphism. The sup is not always assumed. If it is assumed by a unique measure, U is called intrinsically ergodic. If with respect to this measure, it is a K-automorphism, we then call U intrinsically K.

Let $B: X \to X$ and $F: Y \to Y$ be Borel isomorphisms of compact metric spaces with finite topological entropy. Let ψ be a Borel measurable integer-valued function on X and define the skew product:

$$T: X \times Y \to X \times Y$$

$$T(x,y) = (B(x), F^{\psi(x)}(y))$$

So, the base map is B and the fiber maps are selected powers of F.

In this paper we "compute" $h(T)$ in terms of B, F and ψ (Theorem B) and give a condition for intrinsic ergodicity. (Theorem C) Then we show that this yields some new, simple examples of intrinsically ergodic homeomorphisms

(Theorem D, Examples 1.6)

<u>Notation:</u>

i) If α is a finite partition of a space and U a Borel isomorphism, $\alpha_U^+ = \bigvee_{k=1}^{\infty} U_\alpha^k$, $\alpha_U = \bigvee_{-\infty}^{+\infty} U_\alpha^k$

ii) $S_n\psi(x) \equiv \sum_{i=0}^{n-1} \psi(B^i x)$

iii) $\pi_1: X \times Y \to X$, $\pi_2: X \times Y \to Y$ are the natural projections.

iv) All measures are Borel probability measures.

The key result is the following:

<u>Theorem A</u>: Let ν be a B-invariant ergodic measure. Let ψ be ν-integrable. Then

1) $\sup_{\{\pi_1(\mu)=\nu \quad T\mu=\mu\}} h_\mu(T) = h_\nu(B) + h(F)|\int\psi d\nu|$

2) If, in addition, $\int\psi d\nu \neq 0$ and F is intrinsically K then the sup in 1) is acheived uniquely by $\nu \times \mu_F$ where μ_F is the measure of maximal entropy for F.

The basic idea of 1 above is as follows: The measure-theoretic system (B,ν) is a factor (T,μ). Thus, $h_\mu(T) \geq h_\nu(B)$ and the difference is the fiber contribution. Now after n iterates, the fiber map from $\{x\} \times Y$ to $\{B^n x\} \times Y$ is $F^{S_n\psi(x)}$ which, considered as a mapping of Y, has "maximal" entropy $|S_n\psi(x)|h(F)$. Thus, the "Maximal" entropy per unit time is $|\frac{S_n\psi(x)}{n}| \cdot h(F)$ which tends a.e. to $|\int\psi d\nu| \cdot h(F)$ by the ergodic theorem. So, this is the fiber contribution.

There are several ways of making this precise. Consider Pinsker's Formula ([P], [p. 60]) let α,β be paritions and U a measure preserving transformation. Then

$$H_\mu(\alpha \vee \beta \mid (\alpha \vee \beta)^+_U) = H_\mu(\beta \mid \beta^+_U) + H_\mu(\alpha \mid \alpha^+_U \vee \beta_U) \tag{1.1}$$

Since B is ν-ergodic it has a generator β ([Pl, p. 81]) which we can consider either as a partition of X or as a partition of $X \times Y$ (i.e., $\{A \times Y: A \in B\}$). Similarly, if α is a partition of Y we can consider it also as a partition of $X \times Y$. Let μ be a T-invariant measure, with $\pi_1(\mu) = \nu$. Then, as is standard ([R]) $\mu = \int \lambda_x \, d\nu(x)$ where each λ_x is supported on $\{x\} \times Y$ and $T(\lambda_x) = \lambda_{Bx}$. Thus, by (1.1) we have

$$h_\mu(T) = \sup_{\{\alpha\,:\,\text{partition of } Y\}} H_\mu(\alpha \vee \beta \mid (\alpha \vee \beta)^+_T)$$

$$= H_\nu(\beta \mid \beta^+_T) + \sup_\alpha H_\mu(\alpha \mid \alpha^+_T \vee \beta_T) \tag{1.2}$$

$$= h_\nu(B) + \sup_\alpha \int H_{\lambda_x}(\alpha \mid \bigvee_{n=1}^{\infty} F^{S_n\psi(x)}(\alpha))\, d\nu(x)$$

(Note: The integrand is $\lim_{k \to +\infty} H_{\lambda_x}(\alpha \mid \bigvee_{n=1}^{k} F^{S_n\psi(x)}(\alpha))$. In the case that μ is a direct product this fact was originally due to Abramov-Roklin ([A-R]) and Adler ([Ad]).

At least in many cases (e.g., if ψ is bounded) one can show directly by the above and the ergodic theorem that A1 holds. (cf. the formulas [Ab-R, p. 259]). This is how we came upon the result. In section 2 we do exactly this in the case $\int \psi d\nu = 0$. But for the case $\int \psi d\nu \neq 0$ we do something different. We use Abramov's formula ([Ab]) and the techniques of inducing and tower building because this will also provide a proof of A2. The idea is to first assume that ψ has only the values $\{0,1,-1\}$ and in this case induce on an appropriate set, which yields a direct product map which is well understood. In particular, for A2 we use an argument of K. Berg ([Be]). Then we lift the information back up to the map T. We then reduce the general case to the special case by a tower argument. After this work was completed, we learned that the use of these techniques for Skew Products is not new. In particular, Belinskaya ([Bel]) and Newton ([N]) used this method to compute the measure-theoretic entropies of T with respect to direct product measures. So, for theorem A we are mainly glueing together the ideas of [Be], [Bel], and [N].

We mention that if B, F, and ψ are continuous, there is a relative variational principle due to Ledrappier and Walters ([LW]) which implies that the sup in A1 is

$$h_\nu(B) + \int h(T,\pi^{-1}(x))\, d\nu(x)$$

Here $h(T,\pi^{-1}(x))$ is the fiber topological entropy (see [B-2]) and it is an invariant function. Thus, the ergodicity of ν and Theorem A1 imply that for ν-a.e. $x \in X$

$$h(T,\pi^{-1}(x)) = h(F)\left|\int \psi \, d\nu\right|$$

This fact, however, can be verified directly using the definition of $h(T,\pi^{-1}(x))$ and the ergodic theorem.

We mention that the hypotheses of A2 are needed for the following reasons: First, if there were no bias (i.e., if $\int \psi d\nu$ were 0) then the action of the fiber maps would not contribute any entropy; and so only in extreme circumstances (unique ergodicity of F) could one expect the sup in A1 to be acheived uniquely. Secondly, intrinsic ergodicity of F is of course a natural assumption. But one needs more since, for example, if $B = F$ is a zero entropy, uniquely ergodic map and $\psi \equiv 1$ then $T = B \times B$ has several invariant measures which project to the unique B-invariant measure--namely both the diagonal and product measures. This indicates that something stronger is needed. (See Remark 2.4). In particular, the intrinsic K assumption will do.

Now assuming ψ is integrable with respect to each B-invariant measure ν define

$$(1.3) \qquad P_+ = \sup_{\{B\nu=\nu\}} h_\nu(B) + h(F)\int\psi\, d\nu$$

$$(1.4) \qquad P_- = \sup_{\{B\nu=\nu\}} h_\nu(B) - h(F)\int\psi d\nu$$

In the case that B, F, and ψ are continuous, the variational principal of Walters ([W]) asserts that P_+ is the usual topological pressure of $h(F)\psi$ and P_- is the topological pressure of $-h(F)\psi$. As is standard in the theory of topological pressure, a measure which maximizes the sup in (1.3) (resp. (1.4)) is called an equilibrium state for $h(F)\psi$ (resp. $-h(F)\psi$). From Theorem A we shall easily prove:

<u>Theorem B</u>: 1. $h(T) = \max (P_+,P_-)$

2. Assume T has a measure μ of maximal entropy. If $P_+ > P_-$ (resp. $P_- > P_+$) then $\pi_1(\mu)$ is an equilibrium state for $h(F)\psi$ (resp. $-h(F)\psi$). If $P_+ = P_-$ and μ is ergodic, then $\pi_1(\mu)$ is an equilibrium state of either $h(F)\psi$ or $-h(F)\psi$.

<u>Theorem C</u>: Assume that F is intrinsically K and assume that B has unique equilibrium states ν_+ and ν_- for $h(F)\psi$ and $-h(F)\psi$ respectively.

1. If $P_+ \neq P_-$ then T has a unique measure of maximal entropy - either $\nu_+ \times \mu_F$ or $\nu_- \times \mu_F$ depending on which of P_+, P_- is larger.

2) If $P_+ = P_-$ and $\nu_+ \neq \nu_-$ then $\nu_+ \times \mu_F$ and $\nu_- \times \mu_F$ are the only ergodic measures of maximal entropy for T.

3) If $P_+ = P_-$ and $\nu_+ = \nu_-$ then $h(T) = h(B)$ and ν_+ is the unique B-invariant measure of maximal entropy. Also, the measures of maximal entropy for T are exactly those which project to ν_+.

Remarks: 1) A similar result holds in one replaces F by a one-parameter flow $\{F_t\}$ and ψ by a real-valued function. Here, one is skewing into the group $\mathbb{R}$ instead of Z. What happens if one skews into other groups?

2) Note that theorem C contains the fact that the direct product of an intrinsically ergodic (finite) entropy homeomorphism with a intrinsically K (finite entropy) homeomorphism is intrinsically ergodic. (This was essentially proved by Berg [Be] and we use this result in the proof). More generally, one can see that a necessary and sufficient condition for the direct product of two (finite entropy) intrinsically ergodic maps to be intrinsically ergodic is that their Pinsker factors be disjoint in the sense of [F]. (See Remark 2.4). This was suggested by Y. Katznelson.

We are really interested in the case when ψ is continuous, B is an irreducible (i.e., transitive) shift of finite type, and F is an aperiodic (i.e., mixing) shift of finite type. Then the hypotheses of theorem D are satisfied and $h(F)$, P_+, P_-, ν_+, ν_-, and μ_F are all computable by the following proceedure ([K]).

(1.5) First $h(F)$ is the log of the largest eigenvalue λ of the matrix which represents it as a shift of finite type. Since ψ is continuous and integer-valued, it must be locally constant. So, by a standard recoding

(Parry [P2]) we can assume that ψ is constant on the symbols of some representation of B as a shift of finite type. Let A denote the matrix of this representation and let A' be the matrix obtained from A by multiplying each row of A by λ^{ψ_i} (where ψ_i is the value of ψ on the symbol corresponding to the ith row). Then P_+ is the log of the largest eigenvalue μ of A'. And ν_+ is the Markov process defined by the matrix

$$\left(\frac{r_j A'_{ij}}{r_i \mu}\right)$$

where $r = (r_i)$ is a right eigenvector of A' corresponding to μ. Similarly, one computes P_- and ν_- by replacing λ^{ψ_i} by $\lambda^{-\psi_i}$ in the above.

Examples (1.6). Let B and F be the full shift on two symbols {0,1}. As usual B(p.q) denotes the Bernoulli shift of weight (p,q). The following three examples illustrate the three possibilities in theorem C.

1) $\psi(x) = \left\{\begin{matrix} 1 & \text{if} & x_0 = 1 \\ 0 & \text{if} & x_0 = 0 \end{matrix}\right\}$

So, the fiber maps are F and the identity. $P_+ = \log(3)$, $P_- = \log(3/2)$ and so $h(T) = \log(3)$. The unique measure of maximal entropy is $B(\frac{1}{3}, \frac{2}{3}) \times B(\frac{1}{2}, \frac{1}{2})$.

2) $\psi(x) = \left\{\begin{matrix} 1 & \text{if} & x_0 = 1 \\ -1 & \text{if} & x_0 = 0 \end{matrix}\right\}$

Here, the fiber maps are F and F^{-1}. $P_+ = P_- = \log(5/2)$ and so $h(T) = \log(5/2)$. There are exactly two ergodic measures of maximal entropy. $B(\frac{4}{5}, \frac{1}{5}) \times B(\frac{1}{2}, \frac{1}{2})$ and $B(\frac{1}{5}, \frac{4}{5}) \times B(\frac{1}{2}, \frac{1}{2})$.

3) $\psi(x) = \left\{\begin{matrix} 0 & \text{if} & x_0x_1 = 00 \text{ or } 11 \\ -1 & \text{if} & x_0x_1 = 01 \\ 1 & \text{if} & x_0x_1 = 10 \end{matrix}\right\}$

Here, ψ is constant on the 2-block representation of the full 2-shift. The corresponding matrix is

$$\begin{pmatrix} 11 & 00 \\ 00 & 11 \\ 11 & 00 \\ 00 & 11 \end{pmatrix}$$

$P_+ = P_- = \log 2$ and so $h(T) = \log 2$. The measuresof maximal entropy are exactly those which project to $B(\frac{1}{2}, \frac{1}{2})$.

At least for shifts of finite type, cases 2 and 3 are somewhat degenerate. To illustrate this, we first mention that if B is a mixing shift of finite type and $h(F) > 0$ then 3) of theorem C occurs exactly when ψ is cohomologous to 0 i.e., when there is a continuous function $u(x)$ on X such that

$$\psi(x) = u(Bx) - u(x)$$

(see [B1, p. 40])*. It is easy to see that u can be choosen to be integer-valued. In this case the map $(x,y) \to (x, F y^{u(x)})$ is a topological conjugacy between the skew product T and the direct product of B with the identity.

Thus, case 3 is extremely degenerate. Case 2 is also degenerate since, if $A = (a_{ij})$ is the 0-1 matrix of B, it requires that the largest real eigenvalues of $(e^{h(F)\psi_i} a_{ij})$ and $(e^{-h(F)\psi_i} a_{ij})$ be equal. This is some sort of symmetry condition on the function ψ. Thus, generically, $P_+ \neq P_-$ if B is a shift of finite type and F is intrinsically K. In this case, T is intrinsically ergodic.

---

*This is false for rotations of the circle (see [FKS, Cor. 2.3]).

The next result gives some idea of how tight the symmetry condition on ψ actually is in some situations.

Theorem D: Let $n \geq 2$ and let B be the full n-shift. If (i) ψ is defined (i.e., constant) on the symbols $\{0,\ldots,n-1\}$, (ii) F is intrinsically K and (iii) $h(F) \geq \log n$, then T is intrinsically ergodic iff there does not exist a permutation $\sigma: \{1,\ldots,n\} \to \{1,\ldots,n\}$ such that $\psi(i) = -\psi(\sigma(i))$ $i = 1,\ldots,n$.

Proof: By the remarks above, T is intrinsically, ergodic iff $P_+ \neq P_-$. In this case, one sees by (1.5) that (since all the columns sums are the same)

$$P_+ = \sum_{i=1}^{n} e^{h(F)\psi_i}$$

$$P_- = \sum_{i=1}^{n} e^{-h(F)\psi_i}$$

So, it suffices to show that the map

$$\phi: \{(x_1,x_2,\ldots,x_n) \in Z^n: x_1 \geq x_2 \geq \ldots \geq x_n\} \to \mathbb{R}$$

$$\phi(x_1,x_2,\ldots,x_n) = \sum_{i=1}^{n} e^{h(F)x_i}$$

is a 1-1 map. Well, if

$$\phi(x_1,x_2,\ldots,x_n) = \phi(y_1,\ldots,y_n)$$

then

$$e^{h(F)y_1}(e^{h(F)(x_1-y_1)}-1) = \sum_{i=2}^{n} e^{h(F)y_i} - \sum_{i=2}^{n} e^{h(F)x_i}$$

$$< e^{h(F)y_1}(n-1)$$

Thus,

$$e^{h(F)(x_1-y_1)} < n$$

But by assumption $e^{h(F)} \geq n$. Thus, $x_1 \leq y_1$. But by symmetry $y_1 \leq x_1$. So $x_1 = y_1$. Then one sees that we can continue inductively until $x_1 = y_1, x_2 = y_2, \ldots, x_{n-1} = y_{n-1}$ and then it is clear that $x_n = y_n$ also. □

That condition (iii) in theorem D is necessary is illustrated by the example: B = full 4-shift, F = shift of finite type determined by

$$\begin{pmatrix} 1 & 1 & 1 \\ 0 & 0 & 1 \\ 1 & 2 & 1 \end{pmatrix}$$

and ψ assumes the values 2, -1, -1, -1 an the 4 symbols. Here $h(F) < \log 4$, $P_+ = P_-$ and $\nu_+ \neq \nu_-$.

We mention that these considerations also apply to the case when B is a transitive Axiom A basic set, F is a mixing Axiom A basic set and ψ is a piecewise continuous integer-valued function. Moreover, one can use these maps as F and iterate the skewing process to get other examples. Finally, we mention that if F is merely a transitive Axiom A basic set one can get similar results on the number of ergodic measures of maximal entropy.

The motivation for these problems was mainly to find some new simple examples of intrinsically ergodic systems. In particular, the problem was brought to our attention by L. Young, who has computed the topological entropy of another skew-product: the Abraham-Smale example. For C^1 perturbations of this example, topological entropy and conditions for intrinsic ergodicity are obtained in [N-Y].

Conversations with Y. Katznelson, K. Petersen, and B. Weiss were particularly useful in the present formulation and generality of the results.

Section 2: A special case of Theorem A.

In this section, we assume

$$\psi : X \to \{1,0,-1\}$$

Now fix ν, a B-invariant ergodic measure and a ν-integrable ψ. Let $\overline{X} = \{x \in X : \forall n \geq 1 \ \ S_n\psi(x) > 0\}$. If $\nu(\overline{X}) > 0$ then the first return (i.e. induced) map is well-defined a.e.:

$$\overline{B} : \overline{X} \to \overline{X}$$

$$\overline{B}(x) = B^{r(x)}(x)$$

$r(x)$ = time of first return to $\overline{X}$.

Lemma (2.1).* (a) if $\nu(\overline{X}) > 0$ then $\forall x \in \overline{X}$ $S_{r(x)}\psi(x) = 1$

(b) if $\int \psi d\nu \geq 0$ then $\nu(\overline{X}) = \int \psi d\nu$

Proof: (a) Since $x \in \overline{X}$, $\psi(x) = 1$. So, the result is obvious if $r(x) = 1$. Assume now that $r(x) \geq 2$. Since $x \in \overline{X}$, $S_{r(x)}\psi(x) \geq 1$. So it suffices to show

$$(*) \qquad \psi(Bx) + \ldots + \psi(B^{r(x)-1}(x)) \leq 0$$

Now since $B(x) \notin \overline{X}$ $\exists$ k such that

$$\psi(Bx) + \ldots + \psi(B^k x) \leq 0 .$$

But since $B^{r(x)}(x) \in \overline{X}$, it follows that we can find such a $k \leq r(x) - 1$. If $k = r(x) - 1$, we're done. If not, repeat the argument with Bx replaced by $B^{k+1}(x)$. After several repetitions one sees that (*) can be decomposed into finitely many pieces each of which is nonpositive.

*Cf. Marcus-Petersen, these proceedings.

(b) if $\nu(\overline{X}) > 0$ this follows from (a) since

$$\nu(\overline{X}) = \int_{\overline{X}} S_{r(x)}\psi(x)d\nu = \sum_{i=1}^{\infty}\sum_{j=0}^{i-1}\int_{B^j(r^{-1}(i))}\psi d\nu = \int\psi d\nu$$

if $\nu(\overline{X}) = 0$, this follows from the maximal ergodic theorem since

$$0 \leq \int\psi d\nu = \int_{(\overline{X})^c}\psi d\nu \leq 0 .$$

□

Now, we first consider the case $\int\psi d\nu = 0$. Of course, this only pertains to part (1) of Theorem A. By Lemma (2.1b), we have $\nu(\overline{X}) = 0$. But applying the same fact to $-\psi$ we get

$$\nu(\{x \in X : \forall n \geq 1 \; S_n\psi(x) < 0\}) = 0 .$$

Since ψ assumes only the values 0, 1, and -1 it follows that for ν a.e. $x \in X$ there exist $n \geq 1$ with $S_n\psi(x) = 0$. But this means that if α is any partition of $Y, \alpha \subset \bigvee_{n=1}^{+\infty} F^{S_n\psi(x)}(\alpha)$. So by (1.2), the version of Pinsker's formula in section 1, we have that for every T-invariant measure μ with $\pi_1(\mu) = \nu$

$$h_\mu(T) = h_\nu(B)$$

as desired.

Now we consider the case $\int\psi d\nu > 0$ (the case $\int\psi d\nu < 0$ follows from this simply by replacing F by F^{-1}).

By Lemma (2.1b) we have $\nu(\overline{X}) > 0$ and so the induced map $\overline{B}$ is well-defined. Let $\overline{T}$ be the induced map of T on $\overline{X} \times Y$. Then by Lemma 2.1a,

$$\overline{T} = \overline{B} \times F \quad .$$

Let $\overline{\nu} = \dfrac{\nu|\overline{X}}{\nu(\overline{X})}$, the induced measure on $\overline{X}$. In the following $\overline{\mu}$ will denote an arbitrary $\overline{T}$-invariant probability measure on $\overline{X} \times Y$ let $\overline{\pi}_1 : \overline{X} \times Y \to \overline{X}$, $\overline{\pi}_2 : \overline{X} \times Y \to Y$ denote the induced projections.

<u>Lemma 2.2.</u> (a) $\sup_{\{\pi_1(\bar\mu)=\bar\nu\}} (h_{\bar\mu}(\bar T)) = h_{\bar\nu}(\bar B) + h(F)$

(b) if F is intrinsically K then the sup in (1) is achieved uniquely by $\bar\mu = \bar\nu \times \mu_F$.

Proof. (a) By the standard entropy formulas,

$$h_{\bar\mu}(\bar T) = \sup_{\substack{\beta \text{ a partition of } \bar X \\ F \text{ a partition of } Y}} h_{\bar\mu}(\bar T, \beta \times F)$$

$$\leq h_{\bar\nu}(\bar B) + h_{\pi_2(\bar\mu)}(F) .$$

Also, by the addition formula we have equality if $\bar\mu = \bar\nu \times \pi_2(\bar\mu)$. This yields (a).

(b) It follows from the foregoing that $\bar\nu \times \mu_F$ achieves the sup. Now let $\bar\mu$ be a $\bar T$-invariant measure which achieves the sup. Then by the above, $\pi_2(\bar\mu) = \mu_F$. Thus, the measure-theoretic system $(\bar T, \bar\mu)$ has two natural factors $(\bar B, \bar\nu)$ and (F, μ_F). Now both of these are ergodic: $\bar B$ because B is ν-ergodic and inducing preserves ergodicity; F because it is assumed to be intrinsically K. Thus, both factors have generators β and F respectively. [Pl, p. 81]. Now, to say that $\bar\mu$ is the product measure of its marginals $\bar\nu$ and μ_F, is precisely the same as saying that $\beta_{\bar T}$ and $F_{\bar T}$ are independent. (Note that here we are thinking of β and F both as partitions of $X \times Y$ in the natural way.)

To prove this, note that since $\beta \vee F$ is a generator for $\bar T$

$$H_{\bar\mu}(\beta \vee F \mid (\beta \vee F)^+_{\bar T}) = h_{\bar\mu}(\bar T) \tag{2.3}$$

$$= h_{\bar\nu}(\bar B) + h(F)$$

$$= H_{\bar\mu}(\beta|\beta^+_{\bar T}) + H_{\bar\mu}(F|F^+_{\bar T}) .$$

Now for completeness we include the proof (due to Berg ([Be])) that this forces

$\beta_{\overline{T}}$ and $F_{\overline{T}}$ to be independent:

Pinsker's formula (1.1) asserts

$$H_{\overline{\mu}}(\beta \vee F) \mid (\beta \vee F)^{+}_{\overline{T}}) = H_{\overline{\mu}}(\beta|\beta^{+}_{\overline{T}}) + H_{\overline{\mu}}(F|F^{+}_{\overline{T}} \vee \beta_{\overline{T}}).$$

Comparing this with (2.3), we see

$$H_{\overline{\mu}}(F|F^{+}_{\overline{T}}) = H_{\overline{\mu}}(F|F^{+}_{\overline{T}} \vee \beta_{\overline{T}}) .$$

This means that conditioned on $F^{+}_{\overline{T}}$, F is independent of $\beta_{\overline{T}}$. But since independence is a symmetric property, we have $\forall n, m, i$

$$\begin{aligned} H_{\overline{\mu}}(\beta^{n-i}_{m-i} \mid F^{+}_{\overline{T}}) &= H_{\overline{\mu}}(\beta^{n-i}_{m-i} | F^{+}_{\overline{T}} \vee F) \\ &= H_{\overline{\mu}}(\beta^{n-i}_{m-i} \mid F^{+\infty}_{0}) \end{aligned}$$

(where $\beta^{k}_{\ell} = \overline{T}^{\ell}(\beta)\vee\ldots\vee\overline{T}^{k}(\beta)$ etc.). Thus for all i,

$$H_{\overline{\mu}}(\beta^{n}_{m}|F^{+\infty}_{i+1}) = H_{\overline{\mu}}(\beta^{n}_{m}|F^{+\infty}_{i})$$

and so is constant as a function of i. But since F is a K-automorphism, $F^{+\infty}_{i}$ tends to the trivial σ-algebra. Thus, $H_{\overline{\mu}}(\beta^{n}_{m}) = H_{\overline{\mu}}(\beta^{n}_{m}|F^{+\infty}_{i})$ for all i. This means that $\beta_{\overline{T}}$ and $F_{\overline{T}}$ are independent. □

Remark (2.4). In the proof above, the assumption that F be a K-automorphism is not necessary. It would be sufficient to assume (the weaker hypothesis) that the Pinsker factors of F and $\overline{B}$ be disjoint (in the sense of [F]). To see this, note that the the Pinsker factor of $\overline{B}$ would then be independent of the Pinsker factor of F and so in particular would be independent of the tail-field of F (See [P, p. 61]). Thus the proof above shows that the Pinsker factor of $\overline{B}$ would actually be independent of $F_{\overline{T}}$. But then reversing the roles of F and β in the proof above one sees that $F_{\overline{T}}$ would then be independent of $\beta_{\overline{T}}$ as desired. It is not hard to see that this disjointness condition is also necessary.

Now we prove theorem A in the case we are dealing with: $\psi : X \to \{1,0,-1\}$ and $\int \psi d\nu > 0$.

So, let μ be a T-invariant probability measure on $X \times Y$ with $\pi_1(\mu) = \nu$. Let $\overline{\mu} = \dfrac{\mu|\overline{X}\times Y}{\mu(\overline{X}\times Y)}$. By Abramov's Formula ([Ab])

$$\begin{aligned} h_\mu(T) &= h_{\overline{\mu}}(\overline{T})\cdot\mu(\overline{X}\times Y) \\ &= h_{\overline{\mu}}(\overline{T})\cdot\nu(\overline{X}) . \end{aligned}$$

Now since B is ν-ergodic and $\nu(\overline{X}) > 0$ it follows that the map {T-invariant measures on $X \times Y$ with $\pi_1(\mu) = \nu$} $\to$ {$\overline{T}$-invariant measure on $\overline{X} \times Y$ with $\overline{\pi}_1(\overline{\mu}) = \overline{\nu}$} defined by

$$\mu \longrightarrow \overline{\mu}$$

ia a bijection. Thus,

$$\sup_{\pi_1(\mu)=\nu} h_\mu(T) = (\sup_{\overline{\pi}_1(\overline{\mu})=\overline{\nu}} h_{\overline{\mu}}(\overline{T}))\cdot\nu(\overline{X})$$

(by lemma (2.2) a)
$$= (h_{\overline{\nu}}(\overline{B}) + h(F))\cdot\nu(\overline{X})$$
$$= h_\nu(B) + h(F)\int\psi d\nu$$

The latter equality holds by (2.1)b and Abramov's Formula applied to B.

This gives part (1) of Theorem A in this case. As for part (2), if μ maximizes the sup in A1, then $\overline{\mu}$ maximizes $\sup\limits_{\overline{\pi}_1(\overline{\mu})=\overline{\nu}} h_{\overline{\mu}}(\overline{T})$ and so $\overline{\mu} = \overline{\nu} \times \mu_F$ by Lemma (2.2)b. But the map $\mu \longrightarrow \overline{\mu}$ is 1 - 1 and it is easy to see that $\overline{\nu \times \mu_F} = \overline{\nu} \times \mu_F$. □

Section 3: Theorem A: The General Case

We construct a new space $\hat{X}$ for this situation as a tower over X as follows: put a stack of height $|n|$ (including the floor) above the set $\psi^{-1}(n)$ for each $n \neq 0$; the stack above $\psi^{-1}(0)$ is of height 1 i.e. we add nothing at all to $\psi^{-1}(0)$.

$\hat{X}$

Now, since ψ is ν-integrable the measure ν on X naturally defines a probability measure $\hat{\nu}$ on $\hat{X}$ (i.e., push ν up the tower and normalize) $B : X \to X$ naturally defines the tower map $\hat{B} : \hat{X} \to \hat{X}$ which preserves $\hat{\nu}$. Now, define the function $\hat{\psi} : \hat{X} \to \{1,0,-1\}$ by

$$\hat{\psi}(\hat{x},i) = \operatorname{sign}(\psi(x)) .$$

This is set up so that

$$(3.1) \qquad \int \hat{\psi} d\hat{\nu} = \left(\int \psi d\nu\right)\hat{\nu}(X) .$$

Then we define $\hat{T} : \hat{X} \times Y \to \hat{X} \times Y$ by $\hat{T}(\hat{x},y) = (\hat{B}(\hat{x}), F^{\hat{\psi}(\hat{x})}(y))$.

Let $\hat{\pi}_1 : \hat{X} \times Y \to \hat{X}$ denote the natural projection. Now a T-invariant probability measure μ on $X \times Y$ naturally defines a $\hat{T}$-invariant probability measure $\hat{\mu}$ on $\hat{X} \times Y$ with $\hat{\pi}_1(\mu) = \hat{\nu}$. By Abramov's formula and the fact that the induced map of $\hat{T}$ on $X \times Y$ is simply T we have

$$h_\mu(T) = \frac{h_{\hat{\mu}}(\hat{T})}{\hat{\mu}(X \times Y)}$$

$$= \frac{h_{\hat{\mu}}(T)}{\hat{\nu}(X)} .$$

Also, since $\hat{\mu}(X \times Y) = \hat{\nu}(X) > 0$ the map $\mu \longrightarrow \hat{\mu}$ is a bijection between the (T-invariant probability measures with $\pi_1(\mu) = \nu$) and the ($\hat{T}$ -invariant probability

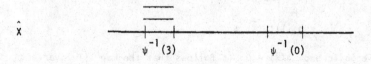

measures with $\hat{\pi}_1(\hat{\mu}) = \hat{\nu}$). Thus,

$$\sup_{(\pi_1(\mu)=\nu)} h_\mu(T) = \sup_{\hat{\pi}_1(\hat{\mu})=\hat{\nu}} \frac{h_{\hat{\mu}}(\hat{T})}{\hat{\nu}(X)}$$

(by the special case) $$= \frac{h_{\hat{\nu}}(\hat{B}) + h(F)|\int\hat{\psi}d\hat{\nu}|}{\hat{\nu}(X)}$$

(by (3.1) and Abramov's formula applied to $\hat{B}$) $$= h_\nu(B) + h(F)\left|\int\psi d\nu\right|.$$

This gives Theorem A1. As for A2, note that if μ maximizes the sup in A1 then $\hat{\mu}$ maximizes $\sup_{\hat{\pi}_1(\hat{\mu})=\hat{\nu}} h_{\hat{\mu}}(T)$ and so by section (2) $\hat{\mu} = \hat{\nu} \times \mu_F$. But since $\mu \longmapsto \hat{\mu}$ is 1 - 1 and $\widehat{\nu \times \mu_F} = \hat{\nu} \times \mu_F$ we have that $\nu \times \mu_F$ is the unique measure which maximizes the sup.

Section 4: Proofs of Theorems B and C

Proof of Theorem B:

(1) By Theorem A1,

$$\sup_{\{\mu:\pi_1(\mu)\text{ ergodic}\}} h_\mu(T) = \sup_{\nu\text{ ergodic}} (h_\nu(B) + h(F)\ |\int\psi d\nu|)$$

$$= \sup_{\nu\text{ ergodic}} (\max(h_\nu(B) + h(F)\int\psi d\nu, h_\nu(B) - h(F)\int\psi d\nu))$$

$$= \max(\sup_{\nu\text{ ergodic}} (h_\nu(B) + h(F)\int\psi d\nu),\ \sup_{\nu\text{ ergodic}} (h_\nu(B) - h(F)\int\psi d\nu))$$

By ergodic decomposition of measures and entropy ([J], [DGS]) it follows that when computing $h(T)$, P_+ and P_- we need only consider ergodic measures. This together with the fact that if μ is T-ergodic then $\pi_1(\mu)$ is B-ergodic, shows that the above simply reads

$$h(T) = \max(P_+, P_-) .$$

(2) This part of Theorem B is evident from the above.

Proof of Theorem C:

(1) Assume for definiteness that $P_+ > P_-$.

$$h_{\nu_+}(B) + h(F)\int\psi d\nu_+ > h_{\nu_-}(B) - h(F)\int\psi d\nu_- \geq h_{\nu_+}(B) - h(F)\int\psi d\nu_+ .$$

(The latter by definition of ν_- as an equilibrium state.) Thus, $\int\psi d\nu_+ > 0$. So theorems A2 and B2 apply to show that $\mu = \nu_+ \times \mu_F$ is the unique measure maximizing entropy.

(2) Since $P_+ = P_-$ we have

$$h_{\nu_+}(B) + h(F)\int\psi d\nu_+ = h_{\nu_-}(B) - h(F)\int\psi d\nu_- \geq h_{\nu_+}(B) - h(F)\int\psi d\nu_+ .$$

Now, if $\int\psi d\nu_+$ were zero then we would have equality above and so by the uniqueness of equilibrium states $\nu_+ = \nu_-$, contrary to assumption. So, $\int\psi d\nu_+ \neq 0$ and so Theorem A2 applies again to show that $\mu = \nu_+ \times \mu_F$ is the unique measure maximizing entropy subject to $\pi_1(\mu) = \nu_+$. Similarly, $\nu_- \times \mu_F$ is the unique measure maximizing entropy subject to $\pi_1(\mu) = \nu_-$. But by theorem B each ergodic measure of maximal entropy must project to ν_+ or ν_-.

(3) In this case we have

$$h\nu_+(B) + h(F)\int\psi d\nu_+ = h_{\nu_+}(B) - h(F)\int\psi d\nu_+$$

and so $h(F)\int\psi d\nu_+ = 0$ which implies

$$h(T) = P_+ = P_- = h_{\nu_+}(B) = h_{\nu_-}(B) \leq h(B) .$$

Thus, $h(T) \leq h(B)$. But since π_1 is a measure-theoretic factor map from (T,μ) to $(B,\pi_1(\mu))$ and since each B-invariant measure $\nu = \pi_1(\nu \times \mu_F)$ we have $h(T) \geq h(B)$. Thus, $h(T) = h(B)$ and $\nu_+ = \nu_-$ is a measure of maximal entropy which is then unique. Thus, by theorem B2, if μ is a T-invariant measure of maximal entropy, then $\pi_1(\mu) = \nu_+$. Conversely, if $\pi_1(\mu) = \nu_+$ then

$$h_\mu(T) \geq h_{\nu_+}(B) = h(B) = h(T)$$

and so $h_\mu(T) = h(T)$.

References

[Ab] Abramov, "The entropy of an induced automorphism," Dokl. Akad. Nauk SSSR 128 (1959), 647-650.

[A-R] Abramov and Roklin, "Entropy of a skew product transformation with invariant measure," AMS Translations, Ser. 2. 48, 255-265.

[Ad] Adler, "A note on the entropy of skew product transformations," Proc. Am. Math. Soc. 4 (1963), 665-669.

[Bel] Belinskaya, R., "Entropy of a piecewise power skew product," Izv. Vyssh. Ucheb. Zaved. Natem. 18 No. 3 (1974), 12-17.

[Be] Berg, K., "Convolution and invariant measures, maximal entropy," Math. Syst. Theory 3 (1969), 146-150.

[B1] Bowen, R., "Equilibrium states and the ergodic theory of Anosov diffeomorphisms," Springer-Verlag Lecture Notes #470 (1975).

[B2] Bowen, R., "Entropy for Group Endormorphisms and Homogeneous spaces," Trans. Amer. Math. Soc. 153 (1971), 401-413.

[DGS] Denker, M.; Grillenseger, C.; and Sigmund, K., "Ergodic Theory on Compact Spaces," Springer-Verlag Lecture Notes #527.

[FKS] Furstenberg, H.; Keynes, H.; and Shapiro, L., "Prime Flows in Topological Dynamics," Israel J. Math. 14 (1973), 26-38.

[F] Furstenberg, H., "Disjointness in Ergodic Theory," Math. Syst. Th. 1(1967), 1-49.

[G] Goodman, T.N.T., "Relating Topological Entropy with Measure-Theoretic Entropy," Bull. London Math. Soc. 3 (1971), 176-180.

[J] Jacobs, K., "Ergodic Decomposition of the Komologorov-Sinai Invariant," Proc. Internat. Sump. in Ergodic Theory, Acad. Press, NY (1963), 173-190, ed. by F. Wright.

[K] Krieger, W., "On the uniqueness of the Equilibrium state," Math. Syst. Theory 8 (1974), 97-104.

[L-W] Ledrappier, F. and Walters, P., "A Relativized Variational Principle," to appear in Bull. London Math Soc.

[N-Y] Newhouse, S. and Young, L., to appear.

[N] Newton, D., "On the entropy of certain classes of skew-product transformations," Proc. Amer. Math. Soc. 21 (1969), 722-726.

[P1] Parry, W., "Entropy and Generators in Ergodic Theory," Math. Lecture Notes Series, Benjamin, New York 1969.

[P2] Parry, W., "Intrinsic Markov Chains," Trans. Amer. Math. Soc. 112 (1964), 55-66

[R] Rohlin, "Lectures on the entropy theory of measure-preserving transformations," Russian Math. Surveys 22 (1967), 1-52.

[W] Walters, P., "A variational principle for the pressure of continuous transformations," Amer. J. Math. 97 (1976), 937-971.

[Y] Young, L., Univ. California, Berkeley, Dissertation.

BALANCING ERGODIC AVERAGES

Brian Marcus[1] and Karl Petersen[2]

1. Background and results.

We consider a (Lebesgue) probability space $(X, \mathcal{B}, \mu)$ and either a single invertible measure-preserving transformation $T : X \to X$ or a one-parameter group $\{T_t : -\infty < t < \infty\}$ of such transformations. In the latter situation, we assume as usual that the map $(x,t) \mapsto T_t x$ is jointly measurable and that the group property $T_{s+t} = T_s T_t$ $(s,t \in \mathbb{R})$ is satisfied. The purpose of this paper is to establish and comment on some fundamental constraints and symmetries (approximately of the kind found in the Recurrence Theorem and the Maximal Ergodic Theorem) that are forced upon a system by the presence of a finite-measure-preserving action. The results are stated quite easily.

Let us consider first the case of a flow $\{T_t : -\infty < t < \infty\}$. For a function $f \in L^1(X, \mathcal{B}, \mu)$, we define the <u>oscillation set</u> of f to be

$$\mathcal{O} = \mathcal{O}(f) = \{x \in X : \text{there are } t, t' > 0 \text{ with } \frac{1}{t}\int_0^t f(T_s x) > 0 \text{ and } \frac{1}{t'}\int_0^{t'} f(T_s x) < 0\}$$

and the <u>crossing set</u> of f to be $\mathcal{C} = \mathcal{C}(f) = \{x \in X : \text{there is } t > 0 \text{ such that } \frac{1}{t}\int_0^t f(T_s x)ds = 0\}$. It is clear that $\mathcal{O} \subset \mathcal{C}$ up to a set of measure 0.

THEOREM A: $\int_{\mathcal{O}} f d\mu = \int_{\mathcal{C}} f d\mu = 0.$

One can best interpret this result by comparing it to the usual Maximal Ergodic Theorem for flows [W], [YK]: If

---

(1) Research supported in part by N.S.F. Grant MCS 78-01244.
(2) Research supported in part by N.S.F. Grant MCS 76-05786.

$$f^*(x) = \sup_{t>0} \frac{1}{t}\int_0^t f(T_s x)ds$$

and $\alpha \in \mathbb{R}$, then

$$\int_{\{f^*>\alpha\}} fd\mu \geq \alpha\mu\{f^* > \alpha\} .$$

In a previous publication [P], it was proved that there is a reverse inequality: If $\alpha > \int fd\mu$ and $\{T_t\}$ is ergodic, then

$$\int_{\{f^*>\alpha\}} fd\mu \leq 4\alpha\mu\{f^* > \alpha\}.$$

Reverse maximal inequalities or converse Dominated Ergodic Theorems had been found earlier by Burkholder [B] for independent identically-distributed random variables, Stein [S] for the real-variable case, Gundy [G] for certain martingales, Ornstein [O] for the discrete ergodic case, and Derriennic [D] for the ratio ergodic case. The matter has also been discussed by Jones [J]. Theorem A applies directly to this situation. For if $\{T_t\}$ is ergodic and $\alpha > \int fd\mu$, let $g = f - \alpha$. Then $\int gd\mu < 0$, so for almost every x a $t > 0$ can be found for which $\frac{1}{t}\int_0^t g(T_s x)ds < 0$. Therefore $O(g) = \{g^* > 0\} = \{f^* > \alpha\}$, and the following result is immediate.

COROLLARY 1: If $\{T_t\}$ is ergodic and $\alpha > \int fd\mu$, then

$$\int_{\{f^*>\alpha\}} fd\mu = \alpha\mu\{f^* > \alpha\} .$$

Thus we actually have equality in the Maximal Ergodic Theorem in this case. This is possible because the parameter is continuous; for a discrete parameter, equality can hold when Range $f \subset \{-1,0,1\}$, in which case again there is a sort of continuity (see the Corollary to Theorem 1, below). (Another remark: Corollary 1 is really a Corollary of Theorem 2, below, and thus does not depend on the Ergodic

Theorem.)

Of course we need not restrict ourselves to crossings of and oscillations about only the level 0; for each $\alpha \in \mathbb{R}$, let

$$\mathcal{O}_\alpha = \mathcal{O}_\alpha(f) = \{x \in X : \text{there are } t, t' > 0 \text{ with } \frac{1}{t}\int_0^t f(T_s x)ds > \alpha$$
$$\text{and } \frac{1}{t'}\int_0^{t'} f(T_s x)ds < \alpha\}$$

and

$$C_\alpha = C_\alpha(f) = \{x \in X : \text{there is } t > 0 \text{ with } \frac{1}{t}\int_0^t f(T_s x)ds = \alpha\}.$$

The first of the following corollaries is immediate, while the second requires a short auxiliary argument.

COROLLARY 2: $\int_{\mathcal{O}_\alpha} f d\mu = \alpha\mu(\mathcal{O}_\alpha)$, $\int_{C_\alpha} f d\mu = \alpha\mu(C_\alpha)$.

COROLLARY 3: If $\mu(C_\alpha) = 1$, then $\alpha = \int f d\mu$. If $\{T_t\}$ is ergodic, then $\mu(C_\alpha) = 1$ if and only if $\alpha = \int f d\mu$.

Let us turn now to the discrete case. Suppose that $T : X \to X$ is a measure-preserving transformation and $f \in L^1(X,\mathcal{B},\mu)$. We consider the partial sums

$$S_n f(x) = \sum_{k=0}^{n-1} f(T^k x) \qquad (n = 1,2,\ldots)$$

together with their extrema

$$S_* f(x) = \inf_{n\geq 1} S_n f(x) \quad \text{and} \quad S^* f(x) = \sup_{n\geq 1} S_n f(x)$$

and sets of constant sign

$$A = \{x \in X : S_n f(x) > 0 \text{ for all } n \geq 1\}$$

and

$$E = \{x \in X : S_n f(x) < 0 \text{ for all } n \geq 1\}.$$

The basic result in this case is the following.

THEOREM B: $\int f d\mu = \int_A S_* f d\mu + \int_E S^* f\, d\mu$.

The relationship between Theorem A and Theorem B can be seen most easily when T and $\{T_t\}$ are ergodic and $\int f d\mu > 0$. Then $\mu(E) = 0$, so Theorem B says that

(1) $$\int_A S_* f d\mu = \int f d\mu .$$

On the other hand, if we let $\underline{A} = \{x \in X : \frac{1}{t}\int_0^t f(T_s x)ds \geq 0 \text{ for all } t > 0\}$, then $\mathcal{O}^c = \underline{A}$ and Theorem A says that

(2) $$\int_{\underline{A}} f d\mu = \int f d\mu;$$

thus in the continuous-parameter case one can find the average of f by summing over only those initial conditions which lie in the set $\underline{A}$, while in the discrete case one must sum not f but $S_* f$.

In the following pages we will prove (1) first, and then (2) will follow by approximation. With the help of J. Neveu, we have also been able to prove (2) directly (see the last part of the paper)*; then (1) follows easily from (2) by considering a flow built under a constant function. Theorems A and B are direct consequences of (2) and (1), respectively, because of the theorems on ergodic decompositions of measure-preserving actions.

2. The discrete case.

In this section we consider a single measure-preserving transformation $T : X \to X$ on a probability space $(X, \mathcal{B}, \mu)$. Recall that for $f \in L^1(X, \mathcal{B}, \mu)$,

---

* We have recently learned that David Engel has independently found a similar short proof.

$$S_n f(x) = \sum_{k=0}^{n-1} f(T^k x) ,$$

$$S_* f(x) = \inf_{n \geq 1} S_n f(x), \quad \text{and}$$

$$A = \{x \in X : S_n f(x) > 0 \quad \text{for} \quad n \geq 1\}.$$

Parallel to the definition made above in the flow case, let

$$\underline{A} = \{x \in X : S_n f(x) \geq 0 \quad \text{for} \quad n \geq 1\} .$$

Then clearly

$$\int_A S_* f d\mu = \int_{\underline{A}} S_* f d\mu ,$$

since $S_* f = 0$ on $\underline{A} \backslash A$.

THEOREM 1: If T is ergodic and $\int f d\mu \geq 0$, then

$$\int_A S_* f d\mu = \int_{\underline{A}} S_* f d\mu = \int f d\mu .$$

<u>Proof</u>: We deal first with the case when $\mu(\underline{A}) > 0$. Then, by ergodicity, for almost every x there is a smallest integer $n(x) \geq 1$ such that $T^{n(x)} x \in \underline{A}$. We claim that

$$S_* f(x) = \sum_{k=0}^{n(x)-1} f(T^k x) . \tag{2.1}$$

Note $T^{n(x)} x \in \underline{A}$, so that

$$\sum_{k=n(x)}^{m} f(T^k x) \geq 0 \quad \text{for} \quad m \geq n(x),$$

and hence

$$S_* f(x) = \sum_{k=0}^{i-1} f(T^k x) \quad \text{for some} \quad i \quad \text{with} \quad 1 \leq i \leq n(x) .$$

Then if $m \geq i$,

$$\sum_{k=i}^{m} f(T^k x) = \sum_{k=0}^{m} f(T^k x) - \sum_{k=0}^{i-1} f(T^k x) \geq 0;$$

this shows that $T^i x \in \underline{A}$, so that $i = n(x)$.

Because T is ergodic and $\mu(\underline{A}) > 0$, we may form the Kakutani tower decomposition of X with respect to $\underline{A}$: if $\underline{A}_n = \{x \in \underline{A} : n(x) = n\}$, then

$$X = \bigcup_{n=1}^{\infty} \bigcup_{i=0}^{n-1} T^i \underline{A}_n$$

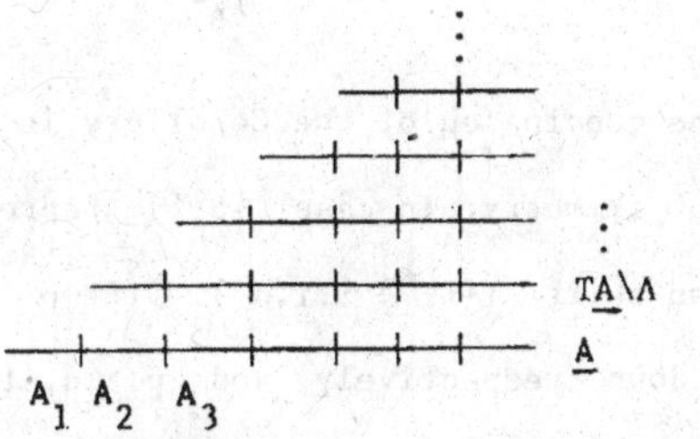

The function f can then be integrated as follows:

$$\int f d\mu = \sum_{n=1}^{\infty} \int_{\underline{A}_n \cup T\underline{A}_n \cup \ldots \cup T^{n-1}\underline{A}_n} f d\mu = \sum_{n=1}^{\infty} \int_{\underline{A}_n} (f + fT + \ldots + fT^{n-1}) d\mu$$

$$= \int_{\underline{A}} \sum_{k=0}^{n(x)-1} f(T^k x) d\mu(x) = \int_{\underline{A}} S_* f d\mu .$$

In case $\mu(\underline{A}) = 0$, we apply (to $-f$) the Maximal Ergodic Theorem for the discrete case: If $g^*(x) = \sup_{n \geq 1} \frac{1}{n} S_n g(x)$, then $\int_{\{g^* > 0\}} g d\mu \geq 0$. The conclusion is that $\int_{\underline{A}^c} f d\mu \leq 0$. Thus

$$0 \leq \int f d\mu = \int_{\underline{A}^c} f d\mu \leq 0,$$

and so $\int f d\mu = 0 = \int_{\underline{A}} S_* f d\mu$, as required.

If f takes only the values -1, 0, and 1, then $S_* f = f$ on A, and this theorem reduces to the following statement.

COROLLARY: If T is ergodic, $\int f d\mu \geq 0$, and Range $f \subset \{-1,0,1\}$, then

$$\mu(A) = \int_A f d\mu = \int f d\mu .$$

Again, the presence of a sort of continuity in this case forces equality to hold in the Maximal Ergodic Theorem:

$$\int_{A^c} f d\mu = 0.$$

The conclusion of the Corollary is familiar, and easy to see from considerations of symmetry, in case $\{S_n f\}$ represents a random walk, i.e. $T : X \to X$ is a two-shift ($\{fT^k\}$ i.i.d.). If p and q are the probabilities of moving up and down, respectively, and $p \geq q$, then the probability of always remaining above the starting position is $\mu(A) = p - q$. It is also familiar for at least some non-stationary processes, for example in the Ballot Problem [F, p. 69]. We see now that such a formula holds for an arbitrary stationary process. This formula, which arose in the first place in connection with certain entropy computations [MN], formed the starting point of our investigations and contains the basic idea of the present paper. Let us sketch two direct proofs of the Corollary in order to illustrate the importance of "0-trains" (which also play a role in the continuous case) in this matter.

<u>First alternative proof</u>: The case when $\mu(A) = 0$ follows again from the Maximal Ergodic Theorem, so suppose that $\mu(A) > 0$. As before, form the Kakutani decomposition of X with respect to A. Then $f(Tx)+\ldots+f(T^{n(x)-1}x) = 0$ for $x \in A_2 \cup A_3 \cup \ldots$. This is so because we can find a smallest $k \geq 1$ for which $f(Tx) + \ldots + f(T^k x) = 0$. Again we must have $k \leq n(x) - 1$. In case $k \neq n(x) - 1$, repeat, with $T^{k+1}x$ in place of x. In this way the sum $f(Tx)+\ldots+f(T^{n(x)-1}x)$ is broken up into a finite number of pieces, each of which sums to 0. Then

$$\int f\,d\mu = \sum_{n=1}^{\infty} \int_{A_n} (f + fT + \dots + fT^{n-1})\,d\mu = \sum_{n=1}^{\infty} \int_{A_n} f\,d\mu = \int_A f\,d\mu.$$

<u>Second alternative proof</u>: In this proof we deal with the possibilities $\mu(A) > 0$ and $\mu(A) = 0$ simultaneously by using Katznelson's and Ornstein's (unpublished) approach to the Maximal Ergodic Theorem. Let

$$B_1 = \{x \in X : f(x) \leq 0\},$$

$$B_2 = \{x \in X : f(x) > 0,\ f(x) + f(Tx) \leq 0\},$$

$$\vdots$$

$$B_n = \{x \in X : f(x) + f(Tx) + \dots + f(T^{n-1}x) \leq 0,\ \sum_{k=0}^{j-1} f(T^k x) > 0 \ \text{ for } \ j < n\}.$$

A moment's thought shows that $T^k B_n \subset B_{n-k} \cup \dots \cup B_1$ for $k = 1,2,\dots,n-1$.

$$\begin{array}{llll} \text{———} \ T^{n-1}B_n & \text{———} \ T^{n-2}B^{(n)}_{n-1} & & \\ \quad\vdots & \quad\vdots & & \\ \text{———} \ TB_n & \text{———} \ TB^{(n)}_{n-1} & & \\ \text{———} \ B_n & \text{———} \ B^{(n)}_{n-1} & \dots & \text{———} \ B^{(n)}_1 \end{array}$$

Picture of $B_1 \cup \dots \cup B_n$.

Let $B^{(n)}_n = B_n$, let $B^{(n)}_{n-1} = B_{n-1} \setminus (B_n \cup TB_n \cup \dots \cup T^{n-1}B_n)$ be the part of B_{n-1} not covered by $B_n \cup TB_n \cup \dots \cup T^{n-1}B_n$,

$$B^{(n)}_{n-2} = B_{n-2} \setminus [(B_n \cup TB_n \cup \dots \cup T^{n-1}B_n) \cup (B^{(n)}_{n-1} \cup TB^{(n)}_{n-1} \cup \dots \cup T^{n-2}B^{(n)}_{n-1})$$

the part of B_{n-2} not previously covered, etc. Then for $k \geq 2$ (by "continuity") $f + fT + \dots + fT^{k-1} = 0$ on $B^{(n)}_k$, and hence

$$\int_{B_1 \cup \dots \cup B_n} f\,d\mu = \sum_{k=1}^{n} \int_{B_k^{(n)}} (f + fT + \dots + fT^{k-1})\,d\mu = \int_{B_1^{(n)}} f\,d\mu = -\mu(B_1^{(n)} \cap f^{-1}\{-1\})$$

(By noting that $\int_{B_k^{(n)}}(f+fT+\ldots+fT^{k-1})d\mu$ is non-positive--whether or not Range $f \subset \{-1,0,1\}$-- and letting $n \to \infty$, one arrives at a proof of the Maximal Ergodic Theorem, $\int_{\{g^*>0\}} g d\mu \geq 0$ (where $g = -f$). Suitably altering the inequalities in the preceding argument shows that also $\int_{\{g^* \geq 0\}} g d\mu \geq 0$. This observation is also a direct consequence of Theorem 1.) Since $A^c = B_1 \cup B_2 \cup \ldots$, in order to show that $\int f d\mu = \int_A f d\mu$ (i.e. $\int_{A^c} f d\mu = 0$), it is enough to show that $\mu(B_1^{(n)} \cap f^{-1}\{-1\}) \to 0$ as $n \to \infty$. This amounts to showing that if $f(x) = -1$, then x is the terminus of some 0-train: i.e., there is $n \geq 1$ with $f(x) + f(T^{-1}x) + \ldots + f(T^{-n}x) = 0$. But if $D = \{x : f(x) + f(T^{-1}x) + \ldots + f(T^{-n+1}x)$ for $n \geq 1\}$, then we must have $\mu(D) = 0$, for otherwise, letting n_D denote the first return time to D (for T^{-1}),

$$\int f d\mu = \int_D \sum_{k=0}^{n_D(x)-1} f(T^{-k}x)d\mu < 0,$$

contrary to assumption.

Finally we remark that Theorem 1 follows easily from its Corollary: one constructs a tower of height $|n|$ on $\{f = n\}$ and replaces f by a function with values in $\{-1,0,1\}$ on the new space.

<u>Proof of Theorem B</u>: According to the theorem on ergodic decompositions (see [AHK]), we may assume that there is a probability space (Ω,F,P) and a family of disjoint T-invariant subsets $\{X_\omega : \omega \in \Omega\}$ of X together with probability measures μ_ω on $\mathcal{B}_\omega = \mathcal{B} \cap X_\omega$ such that

$$X = \bigcup_{\omega\in\Omega} X_\omega,$$

$$\mu(B) = \int_\Omega \mu_\omega(B \cap X_\omega)dP(\omega) \quad \text{for } B \in \mathcal{B},$$

and T is ergodic on $(X_\omega,\mathcal{B}_\omega,\mu_\omega)$ for each $\omega \in \Omega$. By Theorem 1,

$$\int_{A_\omega} S_* f d\mu_\omega = \int_{X_\omega} f d\mu_\omega \quad \text{if} \quad \int_{X_\omega} f d\mu_\omega \geq 0 ;$$

in case $\int_{X_\omega} f d\mu_\omega < 0$, replace f by $-f$ to see that

$$\int_{E_\omega} S^* f d\mu_\omega = \int_{X_\omega} f d\mu_\omega \quad \text{if} \quad \int_{X_\omega} f d\mu_\omega < 0 .$$

Thus

$$\int f d\mu = \int_\Omega \int_{X_\omega} f d\mu_\omega \, dP(\omega) = \int_{\{\omega : \int_{X_\omega} f d\mu_\omega \geq 0\}} \int_{A_\omega} S_* f d\mu_\omega \, dP(\omega) + \int_{\{\omega : \int_{X_\omega} f d\mu_\omega < 0\}} \int_{E_\omega} S^* f d\mu_\omega \, dP(\omega)$$

$$= \int_A S_* f d\mu + \int_E S^* f d\mu .$$

3. The continuous case.

In this section we consider a one-parameter group $\{T_t : -\infty < t < \infty\}$ of measure-preserving transformations on a probability space $(X, \mathcal{B}, \mu)$. For $f \in L^1(X, \mathcal{B}, \mu)$, let

$$F_t(x) = \int_0^t f(T_s x) ds \qquad \text{for} \quad t \geq 0,$$

$A = A(f) = \{x \in X : F_t(x) > 0 \text{ for all } t > 0\}$, and

$\underline{A} = \underline{A}(f) = \{x \in X : F_t(x) \geq 0 \text{ for all } t > 0\}$.

We deal first with a special case of Theorem A.

THEOREM 2: If $\{T_t : -\infty < t < \infty\}$ is ergodic and $\int f d\mu \geq 0$, then

$$\int_A f d\mu = \int_{\underline{A}} f d\mu = \int f d\mu .$$

First a preliminary result is needed.

LEMMA: $\mu\{x \in \underline{A} \setminus A : f(x) > 0\} = 0$.

Proof: By the Local Ergodic Theorem [W], for almost every x with $f(x) > 0$

there is an $\varepsilon(x) > 0$ such that $F_t(x) > 0$ for $0 < t < \varepsilon(x)$. For each $n = 1,2,\ldots$, let

$$E_n = \{x \in \underline{A}\backslash A : f(x) > 0 \text{ and } \varepsilon(x) > \frac{1}{n}\};$$

we will show that $\mu(E_n) = 0$ for all n. For if $\mu(E_n) > 0$, then one may choose t with $0 < t < \frac{1}{n}$ and $\mu(E_n \cap T_{-t}E_n) > 0$ (see [vN]). If $x \in E_n \cap T_{-t}E_n$, then, since $x \in \underline{A}\backslash A$, there is $t_0 > 0$ such that $F_{t_0}(x) = 0$. Therefore

$$t_0 \geq \varepsilon(x) > \frac{1}{n} > t,$$

and

$$0 = F_{t_0}(x) = F_t(x) + F_{t_0-t}(T_t x) .$$

However, this is impossible, because $F_t(x) > 0$ (since $t < \frac{1}{n} < \varepsilon(x)$) and $F_{t_0-t}(T_t x) \geq 0$ (since $t_0 - t > 0$ and $T_t x \in E_n \subset \underline{A}$).

<u>Proof of Theorem 2</u>: Since the Local Ergodic Theorem implies that $f \geq 0$ a.e. on $\underline{A}$, it is immediate from the Lemma that

$$\int_A f d\mu = \int_{\underline{A}} f d\mu .$$

By the continuous-parameter version of the Maximal Ergodic Theorem,

$$\int_{\underline{A}^c} f d\mu \leq 0 .$$

Thus if $\mu(A) = 0$, then

$$0 \leq \int f d\mu = \int_{\underline{A}} f d\mu + \int_{\underline{A}^c} f d\mu \leq \int_{\underline{A}} f d\mu = \int_A f d\mu = 0 .$$

We assume then that $\mu(A) > 0$.

For each $\varepsilon > 0$, let

$$f_\varepsilon(x) = \frac{1}{\varepsilon}\int_0^\varepsilon f(T_s x)ds \quad .$$

It is well known [PS] that for all except possibly countably many $\varepsilon > 0$ the map T_ε is ergodic. Choose a sequence $\varepsilon_1, \varepsilon_2, \ldots$ decreasing to zero with $\varepsilon_k/\varepsilon_{k+1} \in Z$ and each T_{ε_k} ergodic. Let

$$A_k = \{x \in X : \sum_{i=0}^{n-1} f_{\varepsilon_k}(T^i_{\varepsilon_k} x) > 0 \text{ for all } n \geq 1\} \qquad \text{and}$$

$$S_* f_{\varepsilon_k} = \inf_{n\geq 1} \sum_{i=0}^{n-1} f_{\varepsilon_k}(T^i_{\varepsilon_k} x) \quad \text{for } k = 1,2,\ldots \quad .$$

By Theorem 1 and Fubini's Theorem,

$$\int_{A_k} S_* f_{\varepsilon_k} d\mu = \int_X f_{\varepsilon_k} d\mu = \int_X f d\mu \quad \text{for } k = 1,2,\ldots \quad .$$

Now

$$f_\varepsilon(x) + f_\varepsilon(T_\varepsilon x) + \ldots + f_\varepsilon(T_\varepsilon^{n-1} x) = \frac{1}{\varepsilon}\int_0^{n\varepsilon} f(T_s x)ds \ ,$$

$A_k \supset A_{k+1}$ for all k, and

$$A = A(f) \subset \bigcap_{k=1}^{\infty} A_k \subset \underline{A} = \underline{A}(f) \ .$$

Let

$$g_k(x) = \frac{1}{\varepsilon_k}\int_0^{\varepsilon_k} |f(T_s x)| ds \ ,$$

so that $\int g_k d\mu = \int_X |f| d\mu < \infty$ for all k, and $g_k \to |f|$ a.e. as $k \to \infty$. Let

$$h_k = \chi_{A_k} S_* f_{\varepsilon_k} \quad ;$$

then $0 \leq h_k \leq g_k$ a.e. It follows from an extended version of the Dominated Convergence Theorem [R , p. 232] that

$$\int_X h_k d\mu \to \int_X \lim h_k d\mu \ .$$

It is sufficient, then, to prove that

$$h_k \to f \cdot \chi_A \quad \text{a.e.}$$

For this purpose it is enough to show that if $x \in A$ and $f(x) > 0$, then

$$\lim_{\varepsilon \to 0^+} S_* f_\varepsilon(x) = f(x) .$$

For in any case by the Local Ergodic Theorem

$$0 \leq \limsup h_k \leq \chi_{\underline{A}} \cdot f,$$

and so by the Lemma the part of $\underline{A} \setminus A$ where this lim sup could be positive has measure 0.

Suppose then that $x \in A$ and $f(x) > 0$. Because $\mu(A) > 0$ and $\{T_t\}$ is ergodic, we can almost surely find $t_0 > 0$ with $T_{t_0} x \in A$. Now if $n\varepsilon > t_0$, then

$$\int_0^{n\varepsilon} f(T_s x) ds > \int_0^{t_0} f(T_s x) ds > 0 .$$

Since

$$\frac{1}{\varepsilon} \int_0^{\varepsilon} f(T_s x) ds \to f(x) \quad \text{as} \quad \varepsilon \to 0^+ ,$$

for sufficiently small $\varepsilon > 0$

$$\inf_{n \geq 1} \frac{1}{\varepsilon} \int_0^{n\varepsilon} f(T_s x) ds$$

is assumed for some $n \geq 1$ with $n\varepsilon \leq t_0$.

Fix $s_0 > 0$. We claim that for sufficiently small $\varepsilon > 0$, the infimum in the definition

$$S_* f_\varepsilon(x) = \inf_{n \geq 1} \frac{1}{\varepsilon} \int_0^{n\varepsilon} f(T_s x) ds$$

is also not achieved by any n for which $s_0 \leq n\varepsilon \leq t_0$. This is so because $F_t(x)$, a positive continuous function of t, assumes on $[s_0, t_0]$ an absolute minimum value $c > 0$. If our contention were not true, we would have

$$S_* f_\varepsilon(x) \geq \frac{c}{\varepsilon}$$

for arbitrarily small ε, contradicting the fact that

$$0 \leq \limsup_{\varepsilon \to 0^+} S_* f_\varepsilon(x) \leq f(x) \quad \text{a.e. on } A.$$

These comments show that for small $\varepsilon > 0$ we can find a least integer $n_\varepsilon \geq 1$ such that

$$S_* f_\varepsilon(x) = \frac{1}{\varepsilon}\int_0^{\varepsilon n_\varepsilon} f(T_s x)ds ,$$

and that

$$\varepsilon n_\varepsilon \to 0 \text{ as } \varepsilon \to 0^+ .$$

Therefore, for any fixed $\delta > 0$,

$$\frac{1}{\varepsilon n_\varepsilon}\int_0^{\varepsilon n_\varepsilon} f(T_s x)ds > (1 - \delta)f(x)$$

for sufficiently small ε, and hence

$$1 \leq n_\varepsilon < \frac{1}{(1-\delta)f(x)} \frac{1}{\varepsilon}\int_0^{\varepsilon n_\varepsilon} f(T_s x)ds \leq \frac{1}{(1-\delta)f(x)} \frac{1}{\varepsilon}\int_0^{\varepsilon} f(T_s x)ds$$

for sufficiently small ε (by the definition of n_ε). Letting $\varepsilon \to 0$, we see that

$$\lim_{\varepsilon \to 0^+} n_\varepsilon = 1 .$$

Thus $n_\varepsilon = 1$ for small ε, and

$$\lim_{\varepsilon \to 0^+} S_* f_\varepsilon(x) = \lim_{\varepsilon \to 0^+} \frac{1}{\varepsilon}\int_0^{\varepsilon} f(T_s x)ds = f(x) \quad \text{a.e.}$$

Proof of Theorem A: We again use the theorem on ergodic decompositions: we may assume that X is the disjoint union, indexed by a probability space (Ω,F,P), of probability spaces $(X_\omega,\mathcal{B}_\omega,\mu_\omega)$ on each of which $\{T_t : -\infty < t < \infty\}$ acts ergodically.

Fix $\omega \in \Omega$. If $\int_{X_\omega} f d\mu_\omega > 0$, then, with respect to the system $(X_\omega,\mathcal{B}_\omega,\mu_\omega,\{T_t\})$,

$$\mathcal{O}_\omega = \underline{A}_\omega^c \, , \quad \text{while } C_\omega = A_\omega^c$$

(both up to sets of measure 0), by the Ergodic Theorem. Then Theorem 2 implies that

$$\int_{\mathcal{O}_\omega} f d\mu_\omega = \int_{C_\omega} f d\mu_\omega = 0 \; .$$

If $\int_{X_\omega} f d\mu_\omega < 0$, the same conclusion follows from considering $-f$, since $\int_{X_\omega} (-f) d\mu_\omega > 0$, $\mathcal{O}_\omega(-f) = \mathcal{O}_\omega(f)$, and $C_\omega(-f) = C_\omega(f)$. If $\int_{X_\omega} f d\mu_\omega = 0$, then $\int_{X_\omega} (-f) d\mu_\omega = 0$ also, so by Theorem 2

$$\int_{X_\omega \cap \{F_t > 0 \text{ for all } t>0\}} f d\mu_\omega = \int_{X_\omega \cap \{F_t < 0 \text{ for all } t>0\}} f d\mu_\omega = \int_{X_\omega} f d\mu_\omega = 0 \; ;$$

therefore

$$\int_{C_\omega} f d\mu_\omega = 0$$

and, since the Lemma implies that

$$\int_{C_\omega \setminus \mathcal{O}_\omega} f d\mu_\omega = \int_{\underline{A}_\omega(-f) \setminus A_\omega(-f)} f d\mu_\omega + \int_{\underline{A}_\omega(f) \setminus A_\omega(f)} f d\mu_\omega = 0,$$

also

$$\int_{\mathcal{O}_\omega} f d\mu_\omega = 0 \; .$$

Because $\mathcal{O}_\omega = \mathcal{O} \cap X_\omega$ and $C_\omega = C \cap X_\omega$ for each $\omega \in \Omega$, we have that

$$\int_O f d\mu = \int_\Omega \int_{O_\omega} f d\mu_\omega dP(\omega) = 0$$

and

$$\int_C f d\mu = \int_\Omega \int_{C_\omega} f d\mu_\omega dP(\omega) = 0 .$$

Alternative proof of Theorem 2: Suppose that $\{T_t\}$ is ergodic and $f \in L^1(X,\mathcal{B},\mu)$ with $\int f d\mu \geq 0$. If $\mu(A) = 0$, then as before it follows from the Maximal Ergodic Theorem that $\int f d\mu = 0$, as required. We assume then that $\mu(A) > 0$, and prove that

$$\int_{\underline{A}^c} f d\mu = 0 .$$

For fixed $x \in X$, $W_x = \{t \in \mathbb{R} : T_t x \in \underline{A}^c\}$ is an open set. The components of W_x are bounded, since $\mu(A) > 0$ and $\{T_t\}$ is ergodic. Let $(a,b) \subset W_x$ be a component of W_x; we claim that

$$\int_a^b f(T_s x) ds = 0 .$$

(Notice that this proof again depends on finding 0-trains.) Since $a \notin W_x$, $\int_a^t f(T_s x) ds \geq 0$ for all $t \geq 0$. Suppose that $\int_a^b f(T_s x) ds = c > 0$. Choose $t_0 \in (a,b)$ with

$$\int_a^{t_0} f(T_s x) ds < c ,$$

and choose $t_1 \in [t_0, b)$ at which

$$\int_a^t f(T_s x) ds$$

achieves its absolute minimum value on $[t_0, b]$. Then for any $t \geq t_1$, we will have

$$\int_{t_1}^t f(T_s x) ds \geq 0 ,$$

contradicting the fact that $t_1 \in W_x$. Therefore $\int_a^b f(T_s x) ds = 0$.

We will now integrate f over $\underline{A}^c$ by integrating over the orbit pieces, W_x, that lie in $\underline{A}^c$. For each $x \in \underline{A}^c$, let J_x denote the component of W_x which contains 0; if $x \in \underline{A}$, let $J_x = \emptyset$. Define $\lambda : X \to \mathbb{R}$ by

$$\lambda(x) = \begin{cases} 0 & \text{if } x \in \underline{A} \\ \dfrac{1}{\ell(J_x)} & \text{if } x \in \underline{A}^c \end{cases}$$

($\ell(J_x)$ denotes the length of the interval J_x). Since

$$\int_{\{s:-s\in J_x\}} \lambda(T_{-s}x)\,ds = \chi_{\underline{A}^c}(x)$$

and

$$-s \in J_x \text{ if and only if } s \in J_{T_{-s}x} ,$$

applying the measure-preserving change of variables $(x,s) \to (T_s x, s)$ shows that

$$\int_{\underline{A}^c} f(x)d\mu(x) = \int_X f(x)\chi_{\underline{A}^c}(x)d\mu(x) = \int_X f(x)\int_{\{s:-s\in J_x\}} \lambda(T_{-s}x)ds\,d\mu(x)$$

$$= \int_X f(x)\int_{\{s:s\in J_{T_{-s}x}\}} \lambda(T_{-s}x)ds\,d\mu(x) = \int_X \lambda(x)\int_{J_x} f(T_s x)ds\,d\mu(x) = 0 .$$

Of course we could have started with this proof of Theorem 2 and easily deduced from it the other results in this paper. We hope that the alternative arguments we have included are instructive and perhaps offer insights which might be useful for other purposes as well.

REFERENCES

[AHK] Warren Ambrose, Paul R. Halmos, and Shizuo Kakutani, The decomposition of measures II, Duke Math. J. 9 (1942), 43-47.

[B] D. L. Burkholder, Successive conditional expectations of an integrable function, Ann. Math. Stat. 33 (1962), 887-893.

[D] Yves Derriennic, On the integrability of the supremum of ergodic ratios, Ann. of Prob. 1 (1973), 338-340.

[F] William Feller, An Introduction to Probability Theory and its Applications, Vol. I, J. Wiley & Sons, Inc., New York, 1950.

[G] Richard F. Gundy, On the class L log L, martingales, and singular integrals, Studia Math. 33 (1969), 109-118.

[J] Roger L. Jones, Inequalities for the ergodic maximal function, Studia Math. 60 (1977), 111-129.

[MN] B. Marcus and S. Newhouse, Measures of maximal entropy for a class of skew products, to appear.

[O] Donald Ornstein, A remark on the Birkhoff ergodic theorem, Ill. J. Math. 15 (1971), 77-79.

[P] Karl Petersen, The converse of the dominated ergodic theorem, to appear.

[PS] C. Pugh and M. Shub, Ergodic elements of ergodic actions, Compositio Math. 23 (1971), 115-122.

[R] H. L. Royden, Real Analysis (Second Edition), Macmillan Co., New York, 1968.

[S] E. M. Stein, Note on the class L log L, Studia Math. 32 (1969), 305-310.

[vN] J. von Neumann, Über einen Satz von Herrn M. H. Stone, Annals of Math. 33 (1932), 567-574.

[W] Norbert Wiener, The ergodic theorem, Duke Math. J. 5 (1939), 1-18.

[YK] Kôsaku Yosida and Shizuo Kakutani, Birkhoff's ergodic theorem and the maximal ergodic theorem, Proc. Imp. Acad. Tokyo 15 (1939), 165-168.

Mathematics Dept.

University of North Carolina

Chapel Hill, North Carolina, 27514 USA

INVARIANT MEASURES FOR CONTINUOUS TRANSFORMATIONS OF [0,1] WITH ZERO TOPOLOGICAL ENTROPY

by

Michał Misiurewicz

Let $f : [0,1] \longrightarrow [0,1]$ be a continuous transformation with zero topological entropy and let μ be an ergodic f-invariant probability measure on $[0,1]$ which is not concentrated on a periodic orbit of f. We shall give a characterization of the system $([0,1], \mu, f)$.

Take a countable number of copies of the space $\{1,2\}$ with the discrete topology and the measure $\bar{\nu}$ such that $\bar{\nu}(\{1\}) = \bar{\nu}(\{2\}) = \frac{1}{2}$. Let $\Sigma = \prod_1^\infty \{1,2\}$ be the product space with the product topology and the product measure ν. Define a transformation $g : \Sigma \longrightarrow \Sigma$ as follows: $g(x_i)_1^\infty = (y_i)_1^\infty$, where

$$y_i = \begin{cases} 3 - x_i & \text{if } x_j = 1 \text{ for all } j<i , \\ x_i & \text{if } x_j = 2 \text{ for some } j<i \end{cases}$$

(i.e. $g(1,\ldots,1,2,x_k,x_{k+1},\ldots) = (2,\ldots,2,1,x_k,x_{k+1},\ldots)$). g is invertible: $g^{-1}(2,\ldots,2,1,x_k,x_{k+1},\ldots) = (1,\ldots,1,2,x_k,x_{k+1},\ldots)$ and the image of a cylinder is also a cylinder of the same length. Hence g is a homeomorphism and the measure ν is g-invariant.

<u>THEOREM.</u> The systems $([0,1], \mu, f)$ and (Σ, ν, g) are isomorphic.

Let S be the support of μ. We shall use the following properties of S:

(i) S is a closed invariant set

(ii) $f|_S$ is topologically transitive

(iii) S is not finite.

Define:

$S_0 = \{x \in S : fx = x\}$ ((x,fx) is on the diagonal),

$S_1 = \{x \in S : fx > x\}$ ((x,fx) is over the diagonal),

$S_2 = \{x \in S : fx < x\}$ ((x,fx) is under the diagonal).

From (iii) and (ii) it follows that

(iv) S_1 and S_2 are non-empty .

The following lemmata are similar to those of Šarkovskiĭ [3] , but the situation is more complicated because of infiniteness of S. We often make use of the transitivity of $f|_S$ in the following way: For two sets U, V with non-empty (in S) interiors there exists $n>0$ such that $f^nU \cap V \cap S \neq \emptyset$ ("transition from U to V"). Hence, if $U \subset W$ and $V \cap W = \emptyset$, then there exists $x \in W \cap S$ such that $fx \notin W$ ("step from W to $S \setminus W$"). Besides, the proofs are based on the Darboux property of continuous mappings (if an image of an interval contains two points, then it contains the whole interval between them).

<u>Lemma 1.</u> $\sup S_1 \leq \inf (S_0 \cup S_2)$,

$\sup (S_0 \cup S_1) \leq \inf S_2$.

<u>Proof.</u> Denote $p = \inf (S_0 \cup S_2)$. Assume that there exists $q \in S_1$ such that $q > p$. From the elements of $\{x \in S_0 \cup S_1 : x \geq p\}$ we choose r for which fr is the largest. Clearly $r > p$ and $r \in S_0 \cup S_1$.

Suppose first that $r \in S_1$. Then we put $t = \sup \{x < r : fx = x\}$. Consider the set $A = S \cap [t, fr]$.

If there exists $x \in A$ such that $fx > fr$, then $x < fx$, and hence $x \in S_1$. This contradicts the definition of r , because $x \geqslant t \geqslant p$. Therefore for any $x \in A$ we have $fx \leqslant fr$. By the definition of t , every point of $S \cap [t,r]$ belongs to $S_1 \cup S_0$. To make a transition from a neighbourhood of r to a neighbourhood of p we have to make a step from $S \cap [r,fr]$ either to a neighbourhood of p , or at least to the left of t . Since $S \cap [r,fr]$ is closed, there is $u \in S \cap [r,fr]$ such that $fu \leqslant t$ (see Fig. 1).

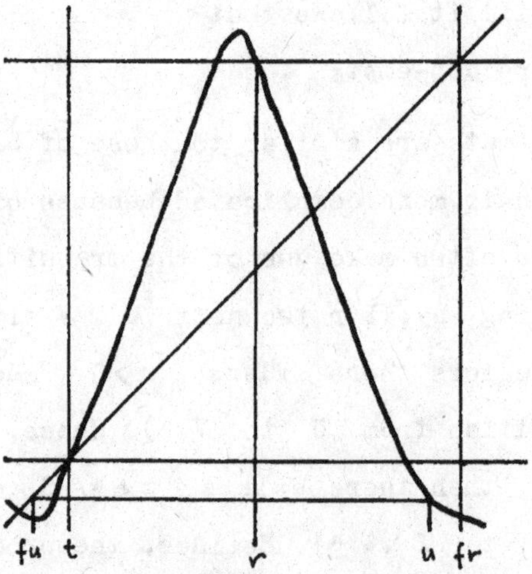

Fig.1.

We obtain a horseshoe effect (see [2]): $f([t,r]) \cap f([r,u]) \supset [t,u]$, and therefore the topological entropy of f $h(f) \geqslant \log 2$ - a contradiction.

Now suppose that $r \in S_0$. From the elements of $\{x \in S_0 \cup S_2 : x \leqslant r\}$ we choose v for which fv is the smallest. If $v \in S_2$, then we obtain for $\bar{f}$ (given by: $\bar{f}x = -f(-x)$) the same situation as earlier, because $S_1(\bar{f}) = -S_2(f)$, $S_2(\bar{f}) = -S_1(f)$, $S_0(\bar{f}) = S_0(f)$, $p(\bar{f}) = -r(f)$, $r(\bar{f}) = -v(f)$. Then $h(f) \geqslant \log 2$ - a contradiction. Hence we have $v \in S_0$. Then $v = p$ and $f(S \cap [p,r]) \subset [p,r]$. From transitivity it follows that $S \subset [p,r]$. Notice that in view of transitivity it is impossible to have $[x,r] \cap S \subset S_1$ for some $x < r$.

Suppose that $[x,r] \cap S \subset S_2$ for some $x < r$. To make a transition from a neighbourhood of p to a neighbourhood of r we have to make a step from $[p,x] \cap S$ to an arbitrarily small neighbourhood of r. Hence, there exists $w \in [p,x] \cap S$ such that $fw = r = fr$ and we can take w instead of r. But $w \in S_1$ and we obtain a contradiction as in the first part of the proof.

Hence, for any $x < r$ in $[x,r]$ there are points of both S_1 and S_2, and thus in $[x,r)$ there are fixed points of f. Take some $y \in S \cap (p,r)$ and a fixed point $z \in (y,r)$. From transitivity it follows that there exists $a \in (p,y) \cap S$ and a positive integer n such that $f^n a \in [z,r]$. Hence, there exists $b \in (p,y)$ such that $f^n b = z$ and thus $f^N b = z$ for any $N \geq n$. But there is a transition in more than n steps from a neighbourhood $(b,z) \cap S$ of y to a neighbourhood $[p,b) \cap S$ of p, and thus there exists $c \in (b,z)$ such that $f^N c < b$ (see Fig. 2 for the graph of f^N).

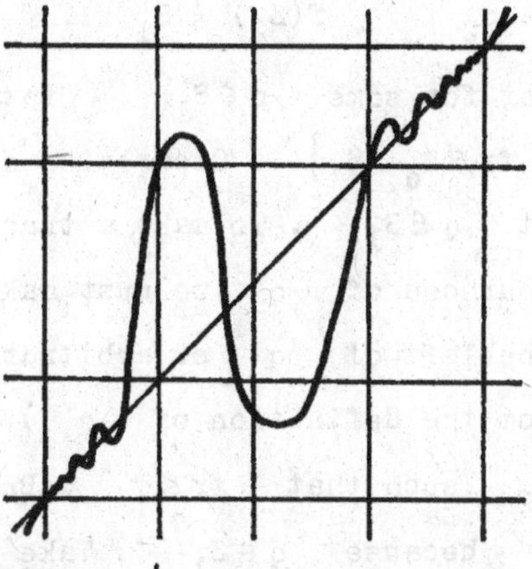

Fig. 2.

We obtain a horseshoe effect for f^N :

$f^N([b,c]) \cap f^N([c,z]) \supset [b,z]$, and therefore

$h(f) = \frac{1}{N} h(f^N) \geqslant \frac{1}{N} \log 2$ - a contradiction.

Hence, we have $\sup S_1 \leqslant \inf (S_0 \cup S_2)$. The inequality $\sup (S_0 \cup S_1) \leqslant \inf S_2$ can be obtained by considering $\bar{f}$ instead of f . ■

<u>Lemma 2.</u> S_0 is either empty or it consists of one point $\sup S_1 = \inf S_2$.

<u>Proof.</u> From Lemma 1 it follows that S_0 is contained in the interval $[\sup S_1 , \inf S_2]$. Thus transitivity implies that S_0 consists of at most two points: $p = \sup S_1$ and $q = \inf S_2$. Suppose that $p \in S_0$ and $p \neq q$. For some small neighbourhood U of p for any $x \in U \cap S$ we have: $x < p$, $fx < q$, and thus $fx \leqslant p$. Since $x \in S_1$, $fx > x$. Therefore all points of $S \cap U$ are contracted towards p and there is no transition from $U \cap S$ to a neighbourhood of q - a contradiction.

If $p \neq q$ and $q \in S_0$, then we consider $\bar{f}$ instead of f . ■

<u>Lemma 3.</u> $f(S_1) = S_2$, $f(S_2) = S_1$.

<u>Proof.</u> Assume that for some $p \in S_1$, $fp \in S_0 \cup S_1$. Take $q = \sup \{x \in S_1 : fx \in S_0 \cup S_1\}$. Clearly, $q \in S_0 \cup S_1$.

Suppose first that $q \in S_1$. To make a transition from S_2 to a neighbourhood of q we must make a step from S_2 either to the left of q or arbitrarily close to q (it follows from the definition of q). Hence, there exists $r \in S_0 \cup S_2$ such that $fr \leqslant q$. But r cannot belong to S_0 because $q \in S_1$. Take

$t = \inf \{x \in S_2 : fx \leqslant q\}$. Clearly, t also belongs to S_2 . To make a transition from a neighbourhood of fq to a neighbourhood of t we must make a step from $[q,t) \cap S$ to the right of t or arbitrarily close to t . Hence, there exists $u \in [q,t] \cap S$ such that $fu \geqslant t$. Clearly, $u \in S_1$. There exists a fixed point v such that $\sup S_1 \leqslant v \leqslant \inf S_2$ (see Fig. 3).

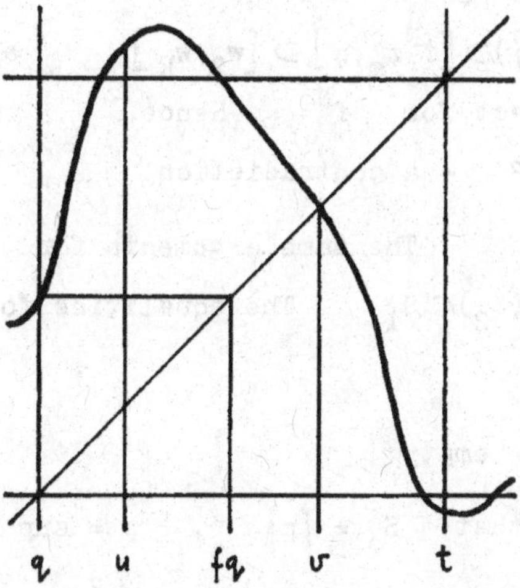

Fig. 3.

We have $fq \leqslant v$, $u \leqslant v$, and hence $f([q,u]) \cap f([u,v]) \supset [u,t]$ and $f([v,t]) \supset [q,v]$. Therefore for f^2 we obtain a horseshoe effect: $f^2([q,u]) \cap f^2([u,v]) \supset [q,v]$, and thus $h(f) = \frac{1}{2} h(f^2) \geqslant \frac{1}{2} \log 2$ - a contradiction.

Suppose now that $q \in S_0$. Then there exists an increasing sequence (w_n) of points of S_1 , converging to q , and such that $fw_n \in S_0 \cup S_1$ for any n . Hence, $fw_n \leqslant q$ for any n . To make a transition from a neighbourhood of w_n to a neighbourhood of w_{n-1} , we have to make a step from $[w_n, q] \cap S$ to S_2 , and therefore there exists

$z_n \in S_1$ such that $z_n > w_n$ and $fz_n > q$. The sequence (z_n) is convergent to q , and thus the sequence (fz_n) is convergent to $fq = q$. Take k such that $z_2 < w_k$ and m such that $fz_m < fz_2$. There is a transition from a neighbourhood of fz_m to a neighbourhood of w_1 , and hence $f^N z_m < w_2$ for some N . But the points q and fz_m belong to $f([w_2,z_2]) \cap f([z_2,w_k])$, and thus $f^N([w_2,z_2]) \cap f^N([z_2,w_k]) \supset [f^N z_m, q] \supset [w_2,w_k]$ and we obtain a horseshoe effect for f^N . Hence, $h(f) = \frac{1}{N} h(f^N) \geqslant \frac{1}{N} \log 2$ - a contradiction.

Thus, $f(S_1) \subset S_2$. The same arguments for $\bar{f}$ instead of f show that $f(S_2) \subset S_1$. The equalities follow from transitivity. ■

Lemma 4. S_0 is empty.

Proof. Suppose that $S_0 = \{p\}$, $p = \sup S_1 = \inf S_2$. From Lemma 3 it follows that $S_0 \cup S_1$ is f^2-invariant. Since $f|_S$ is topologically transitive, so is $f^2|_{S_0 \cup S_1}$. We apply Lemma 1 to f^2 and $S_0 \cup S_1$ instead of f and S , and since p is also a fixed point for f^2 , there is no $x \in S_0 \cup S_1$ for which $f^2 x < x$ - a contradiction. Therefore, by Lemma 2, S_0 is empty. ■

Proof of Theorem. We define by induction:

$S^0 = S$, $S_i^0 = S_i$ $(i = 1,2)$;

$$S^n = S_1^{n-1} , \quad S_1^n = \{x \in S^n : f^{2^n} x > x\} ,$$

$$S_2^n = \{x \in S^n : f^{2^n} x < x\} .$$

Inductive use of Lemmata 3 and 4 shows that:

$S_1^n \cup S_2^n = S^n$, the sets S_i^n are closed, S^n is f^{2^n}-invariant, and $f^{2^n}\Big|_{S^n}$ is topologically transitive

Notice that for given n the sets $f^j(S^n)$, $j = 0,1,\dots,2^n-1$, form a partition of S and $\mu(f^j(S^n)) = 2^{-n}$. We define a mapping $\pi : S \longrightarrow \Sigma$ as follows: $\pi x \in g^j(\{(y_k)_1^\infty : y_k = 1 \text{ for } k = 1,\dots,n\})$ iff $x \in f^j(S^n)$. It is easy to verify, that π is well-defined, continuous, onto, the image of the measure μ under π is equal to ν , and the diagram

$$\begin{array}{ccc} S & \xrightarrow{\;f\;} & S \\ \downarrow{\scriptstyle \pi} & & \downarrow{\scriptstyle \pi} \\ \Sigma & \xrightarrow{\;g\;} & \Sigma \end{array}$$

commutes.

It remains to prove that π is one-to-one μ-almost everywhere. But an inverse image under π of a point is an intersection of S and a family of intervals and thus it is either a single point or an intersection of S with some interval. Since the measure ν is non-atomic, measure μ of such an interval is zero (and consequently, its interior is disjoint from S). All such minimal intervals are pairwise disjoint, and hence there is at most countable number of them. Thus, the measure of their union is zero. ■

<u>Remark 1.</u> We have proved that the mapping π is one-to-one outside of at most countable set, on which it is two- to-one.

<u>Remark 2.</u> If S is finite, then our construction can be continued up to the moment when S^n consists of one point. Thus, our arguments show that if $h(f) = 0$, then any periodic orbit has a period which is a power of 2 (the result of Bowen and Franks [1]).

<u>Remark 3.</u> From the proofs of Lemmata 1 and 3 it follows, that if: $h(f)<\frac{1}{2}\log 2$, f has finitely many local extrema and finitely many fixed points, then any ergodic f-invariant probabilistic measure which is also ergodic with respect to f^2 , is concentrated on a single point.

<u>References.</u>

[1] R. Bowen, J. Franks - The periodic points of maps of the disk and the interval - Topology 15 (1976), 337 - 342

[2] M. Misiurewicz, W. Szlenk - Entropy of piecewise monotone mappings - Asterisque 50 (1977), 299 - 310 (full version will appear in Studia Math. 67)

[3] A.N. Šarkovskiĭ - Coexistence of cycles of a continuous map of a line into itself - Ukr. Mat. Žurnal 16 (1964):1 , 61 - 71 (in Russian)

Institute of Mathematics,
Warsaw University

PKiN, IXp.

<u>00-901 WARSZAWA</u>

POLAND

Dynamical systems of total orders

Jean Moulin Ollagnier

Didier Pinchon

In this short paper, we present some examples of dynamical systems of total orders in connexion with amenability problems.

First of all, we introduce the compact metrizable space T of all total orders on a countable group. The group G acts in a natural way on T by homeomorphisms. An ergodic minimax theorem in this dynamical system leads to the existence of an ameaning filter for a group with the fixed point property (Følner boxes in the ergodic folkore).

Among all G-invariant probability measures on T, one can select a special one π which is invariant under a wider group of homeomorphisms. This allows to obtain local results from mean results in the theory of information gain, hence giving a characterization of equilibrium measures in statistical mechanics.

We then consider the G-invariant subspace T_0 of T consisting of all total orders isomorphic to Z. The existence of an invariant Borel probability on this standard (non compact) space implies the amenability of the group.

One can indeed in this case, not only prove the fixed point property, but directly show the existence of an ameaning filter.

Total orders were first introduced in ergodic theory by Kieffer to show the L^1 convergence of the mean information.

1- The dynamical system (T,G).

a- The compact space T.

Let T be the set of all total orders on the set G. For a finite part F of G and a given total order t on F, we denote by O(F,t) the subset of T consisting of all total orders, the restriction of which to F is t. The set T, endowed with the topology generated by the O(F,t) is a compact, totally disconnected Hausdorff space. Moreover, when G is countable, T is metrizable.

b- The probability π.

Let B denote the group of all permutations of the set G and B_f the invariant subgroup of finite permutations (only moving a finite number of points in G).
Let us denote by b the homeomorphism of the compact set T induced by a given element b of B in the following way

$$x \, \tau \, y \iff b(x) \, b(\tau) \, b(y)$$

i.e. $$x \, b(\tau) \, y \iff b^{-1}(x) \, \tau \, b^{-1}(y)$$

One can easily verify that there exists one and only one Radon probability measure on T, invariant under all the homeomorphisms b induced by permutations of G. This probability gives the measure $1/|F|!$ to $O(F,t)$.
Right and left tranlations on G are permutations, hence proving the existence of an invariant probability for the dynamical system (X,T) without any assumption of amenability on the group G.

c- The cone $\Sigma(G)$.

Let us consider functions f from the set F(G) of all finite parts of G to R verifying the four following conditions

(1) $f(\emptyset) = 0$

(2) $f(Aa) = f(A)$ invariance under right translations

(3) $f(A \cup B) + f(A \cap B) \leqslant f(A) + f(B)$ strong subadditivity

(4) $\exists K > 0 , \forall a \in G , \forall A \in F(G),$

$f(A \cup a) - f(A) \geqslant -K$

The set of all such functions is a convex cone $\Sigma(G)$, on which we define a function q by

$$q(f) = \inf_{A \neq \emptyset} |A|^{-1} f(A)$$

The existence of the ameaning filter for G is nothing else than the result $q(m_D) = 0$ for every finite part D of G containing e where the element m_D of $\Sigma(G)$ is defined by

$$m_D(A) = \left|\{x \in A, \exists d \in D, dx \notin A\}\right|$$

For details, see [2].

d- A minimax theorem in (T,G).

Let f belong to $\Sigma(G)$ and a belong to G. The function f_a defined by $f_a(A) = f(A \cup a) - f(A)$ is decreasing because of the strong subadditivity of f and therefore can be extended by monotonicity in an u.s.c. function on $\mathcal{P}(G)$ also denoted by f_a. Moreover, because of condition (4), f_a is bounded below.

We denote once more by f_a the u.s.c. function on T :

$f_a(\tau) = f_a(a_\tau^-)$ where a_τ^- stands for $\{x \in G, x\tau a, x \neq a\}$

When $f = m_D$, we denote the increment $(m_D)_e$ by i_D .

When the group G has the fixed point property, the following minimax theorem holds

Theorem Let f belong to $\Sigma(G)$ and a belong to G.

$$q(f) = \sup \lambda(f_a)$$

where the supremum is taken over all probabilities on T, invariant under the homeomorphisms of T induced by right tranlations of G.

For the proof, see [2].

e- Invariant measures of the i_D.

Let us recall that we denote by i_D the u.s.c. function $(m_D)_e$. In fact i_D is more regular. This function is continuous on T since it only depends on the restriction of the total order τ to a finite part in the following explicit manner

$$i_D(\tau) = 1 - 1_{(e = \sup D)} - \sum_{d \in D-e} 1_{(e = \sup Dd^{-1})}$$

where $(e = \sup E)$ denotes the subset of T : $\{\forall d \in E,\ d \,\tau\, e\}$

For every probability λ on T, invariant under homeomorphisms induced by right translations, we get

$$\lambda(i_D) = 1 - \lambda(e = \sup D) - \sum_{d \in D-e} \lambda(d = \sup D)$$

$\lambda(i_D) = 0$ since $\{(d = \sup D),\ d \in D\}$ is a partition of T.

Therefore, in the case where G has the fixed point property, $q(m_D) = \sup \lambda(i_D) = 0$ hence showing the existence of an ameaning filter for G.

Let us remark that the fixed point property is neither used for showing the existence of an invariant probability on T nor for proving that $\lambda(i_D) = 0$ for every invariant .

The crucial use of the fixed point property is the minimax ergodic theorem, which is a kind of invariant Hahn-Banach theorem.

2- A lemma about amenability.

To characterize equilibrium measures in statistical mechanics as invariant Gibbs measures, it is necessary to obtain local results from mean results. See [3].

The following lemma, using the probability π on T, allows such an operation.

Lemma Let f be a positive, increasing, right invariant function on the set F(G) of all finite parts of G so that

$$\lim_{\mathcal{M}} |A|^{-1} f(A) = 0$$

where $\mathcal{M}$ is the ameaning filter on F(G).

Then, for an arbitrary given x in G,

$$\inf_{A \to G-x} \lim \; (f(A \cup x) - f(A)) = 0$$

Proof. There is no restriction to suppose $f(\emptyset) = 0$, so that we make this assumption.

For every A in F(G) and every total order t on A :

$$f(A) = \sum_{x \in A} f(x \cup x_t^-) - f(x_t^-)$$

whence

$$f(A) = 1/|A|! \sum_t \sum_{x \in A} f(x \cup x_t^-) - f(x_t^-)$$

The function f_x defined on F(G) by

$$f_x(A) = \inf_{A' \supset A} f(A' \cup x) - f(A')$$

is an increasing function of A.

It is then possible to extend f_x to an increasing positive

Borel function on $\mathcal{P}(G)$ and then to a positive Borel function on T by $f_x(\tau) = f_x(x_\tau^-)$.

One can easily verify that

$$f(A) \geq \sum_{x \in A} d\pi(\tau)\ f_x(\tau) = |A| \int d\pi(\tau)\ f_e(\tau)$$

because of the invariance of π.

Then, thanks to the assumption $\lim_{\mathcal{M}} |A|^{-1} f(A) = 0$, there comes

$$\int d\pi(\tau)\ f_x(\tau) = 0$$

If we call ν the probability measure on $\mathcal{P}(G-x)$, image of π by the application $\tau \longrightarrow x_\tau^-$, we get

$$\int d\nu(A)\ f_x(A) = 0$$

f_x is then a positive, increasing function of A in $\mathcal{P}(G-x)$, so we have necessarily

$$\inf \{ f(A' \cup x) - f(A')\ ,\ A' \in F(G-x)\ ,\ A' \supset A \} = 0$$

since ν give to the cylinder $A' \supset A$ of $\mathcal{P}(G-x)$ the measure $1/(|A| + 1)$.

Whence the result

$$\inf \lim_{A \to G-x} [f(A \cup x) - f(A)] = 0$$

3- The dynamical system $(T_\circ, G)$.

Let us consider the subset $T_\circ$ of T consisting of all total orders isomorphic to the order of Z when G is countable.
It is rather simple to verify $\pi(T_\circ) = 0$ since, for total orders in $T_\circ$, given two elements of G, a and b, there are only a finite number of elements between a and b.

$$T_\circ \subset \bigcup_{D \in F(G)} \{ \tau \, , \forall x \notin D \, , \; x \;\tau\; \inf(a,b) \text{ or } \sup(a,b) \;\tau\; x \} .$$

Let X be the set of all bijections from Z to G sending O on the unit element e of G. For a given x in X, $\bar{x}$ denotes the inverse map. X can be considered as a subset of G^Z and also as a subset of Z^G.
Setting on G^Z and Z^G the product topologies of discrete topologies, we make these sets become Polish spaces.
X is then the closed subspace of $G^Z \times Z^G$ defined by

$$X = \{ (x,y) \in G^Z \times Z^G \, , \; x \circ y = Id_G \, , \; y \circ x = Id_Z \, , \; x(0) = e \}$$

and so X is a Polish space.

It is possible to define an action of G on X by homeomorphisms, $x \longmapsto h(x)$, where h(x) is the element of X defined by

$$\forall m \in Z \, , \qquad h(x)(m) = x(m + \bar{x}(h)).h^{-1}$$

Let us consider the map τ from B to T

$$a \;\tau(x)\; b \iff \bar{x}(a) \leq \bar{x}(b) \quad \text{for the usual order on Z.}$$

This map is a continuous injection the image of which is $T_\circ$. Moreover, we have

$$\tau(h(x)) = h(\tau(x))$$

if we denote by h the homeomorphism of T induced by the right translation by h^{-1}.

The map τ is not an homeomorphism since $T_\circ$ is not a Polish space. One can indeed verify that $T_\circ$ does not possess the Baire property. Nevertheless, the Borel σ-algebras are the same since X is Polish.

So we can consider the standard dynamical systems (X,G) or $(T_\circ, G)$.

<u>Theorem</u> If there exists a Borel probability measure P on X, invariant by the action of G, then G is amenable.

Proof. It is sufficient to prove the existence of an ameaning filter, to show that q on $\Sigma(G)$ verifies

$$q(f) = \int d\pi(\tau)\ f_e(\tau)$$

Using P we obtain a linear map $f \to \varphi(f)$ from $\Sigma(G)$ to $\Sigma(Z)$

$$\varphi(A) = \int dP(x)\ f(x(A))$$

We have

$$q(\varphi) \geqslant q(f) \geqslant \pi(f) = \int_{T(G)} d\pi(\tau)\ f_e(\tau)$$

Since Z is amenable

$$q(\varphi) = \pi(\varphi) = \int_{T(Z)} d\pi(\tau)\ \varphi_0(\tau)$$

To achieve the proof, it remains to show $\pi(f) = \pi(\varphi)$.

$$\varphi_0(O_\tau^-) = \int dP(x)\ f_e(x(O_\tau^-))$$

Whence, by using Fubini's theorem

$$\pi(\varphi) = \int_{T(Z)} d\pi(\tau) \int_X dP(x)\ f_e(x(O_\tau^-)) =$$

$$\int_X dP(x) \int_{T(Z)} d\pi(\tau)\ f_e(x(O_\tau^-))$$

The image of π on T(Z) by an x is π on T(G) and

$$\pi(\varphi) = \int_X dP(x) \int_{T(G)} d\pi(\tau)\ f_e(e_\tau^-) = \pi(f).$$

Remark. It is easy to see that the existence of an hyperfinite action of G on a probability space implies the existence of P on X.

References

[1] KIEFFER J.C. A generalized Shannon-McMillan theorem for the action of an amenable group on a probability space. Annals of Proba. Vol.3 Nb 6 (1975) 1031-1037.

[2] MOULIN OLLAGNIER J. et PINCHON D. Une nouvelle démonstration du théorème de E.Følner. C.R.A.S. 287 (1978)

[3] ———— Riemann integration and the variational principle for amenable groups. Preprint 1978.

J.M.O. Département de Mathématiques, Université Paris-Nord, Avenue J.B.Clément F 93430 VILLETANEUSE

D.P. Laboratoire de Probabilité, Université P.et M. Curie, 4, place Jussieu F 75230 PARIS CEDEX 05

An information obstruction to finite expected coding length

by

William Parry

Ever since Ornstein proved his isomorphism theorem for Bernoulli automorphisms with equal entropy it has been natural to enquire into the possibility of improvements which, for example, would take into account the 'states' of the automorphisms. Now that Keane and Smorodinsky [K.S.], have proved that finitary isomorphisms exist for Bernoulli automorphisms with equal entropy one wonders how reasonable such isomorphisms might be. The purpose of this note is to show there are pairs of Bernoulli automorphisms with equal entropy having the property that whatever finitary isomorphism is chosen it will have the drawback that either it, or its inverse, will have infinite expected coding length.

§1. Preliminaries

We shall assume that readers are familiar with basic definitions and properties of entropy theory. In particular, for a probability space $(X,\mathcal{B},m)$ with sub-σ-algebras $\mathcal{A}_1, \mathcal{A}_2, \mathcal{C}$ we have the basic identity for conditional information:

$$(1.1)\quad I(\mathcal{A}_1 \vee \mathcal{A}_2 | \mathcal{C}) = I(\mathcal{A}_1|\mathcal{C}) + I(\mathcal{A}_2|\mathcal{A}_1 \vee \mathcal{C}).$$

If $\alpha = (A_1,..A_n)$ is a finite partition and $\mathcal{C}$ is a sub-σ-algebra we define

$$d(\alpha,\mathcal{C}) = \inf\{ \sum_{i=1}^{n} m(A_i \Delta C_i) : \gamma = (C_1,\ldots,C_n),\ \gamma \subset \mathcal{C}\},$$

and for sub-σ-algebras $\mathcal{A}, \mathcal{C}$ we define

$$d(\mathcal{A}, \mathcal{C}) = \sup \{d(\alpha, \mathcal{C}) : \alpha \subset \mathcal{A}\} \text{ and}$$

$$D(\mathcal{A}, \mathcal{C}) = \max [d(\mathcal{A}, \mathcal{C}), d(\mathcal{C}, \mathcal{A})].$$

The following properties of d are proved in a routine fashion:

(1.2) $0 \leq d(\mathcal{A}, \mathcal{C}) \leq 2$ with $d(\mathcal{A}, \mathcal{C}) = 0$ when and only when $\mathcal{A} \subset \mathcal{C}$,

(1.3) $d(\mathcal{A}_1, \mathcal{A}_3) \leq d(\mathcal{A}_1, \mathcal{A}_2) + d(\mathcal{A}_2, \mathcal{A}_3)$,

(1.4) $d(\mathcal{A}_1, \mathcal{A}_3) \leq d(\mathcal{A}_1, \mathcal{A}_2)$ when $\mathcal{A}_2 \subset \mathcal{A}_3$

(1.5) $d(\mathcal{A}_1, \mathcal{A}_3) \leq d(\mathcal{A}_2, \mathcal{A}_3)$ when $\mathcal{A}_1 \subset \mathcal{A}_2$;

if $\mathcal{A}_n \uparrow \mathcal{A}$ then $d(\mathcal{A}, \mathcal{C}) = \lim_{n\to\infty} d(\mathcal{A}_n, \mathcal{C})$ and

(1.6) $d(\mathcal{C}, \mathcal{A}) = \lim_{n\to\infty} d(\mathcal{C}, \mathcal{A}_n)$.

If T is an automorphism then

(1.7) $d(\mathcal{A}, \mathcal{C}) = d(T^{-1}\mathcal{A}, T^{-1}\mathcal{C})$.

(1.8) $d(\mathcal{A}_1 \vee \mathcal{A}_2, \mathcal{C}_1 \vee \mathcal{C}_2) \leq d(\mathcal{A}_1, \mathcal{C}_1) + d(\mathcal{A}_2, \mathcal{C}_2)$ and therefore

(1.9) $d(\mathcal{A}_1 \vee \mathcal{A}_2, \mathcal{C}) = d(\mathcal{A}_1, \mathcal{C})$ when $\mathcal{A}_2 \subset \mathcal{C}$.

Clearly D is a metric on the set of all sub-σ-algebras.

Our basic observation is the following

Lemma 1. [P.1]. If $\mathcal{A}, \mathcal{C}$ are two sub-σ-algebras then $d(\mathcal{A}, \mathcal{C}) < 2$ if and only if $I(\mathcal{A} | \mathcal{C}) < \infty$ on a set of positive measure.

A formal proof of this can be given (cf. [P.1]) but we shall provide a heuristic argument instead. We limit ourselves to Lebesgue spaces $(X, \mathcal{B}, m)$. Evidently there is no loss in generality in assuming that $\mathcal{A} = \mathcal{B}$. Consider the measurable partition γ corresponding to $\mathcal{C}$. This fibres the space X and we consider measurable cross sections of positive measure, if they exist. In this case it is not difficult to convince oneself that $d(\mathcal{A}, \mathcal{C}) < 2$ and

and $I(\mathfrak{a}|\mathcal{C})<\infty$ (on a set of positive measures). If such cross sections do not exist then X may be represented as a direct product space with γ as vertical fibres and with vertical space non-atomic. In this case $d(\mathfrak{a},\mathcal{C}) = 2$ and $I(\mathfrak{a}|\mathcal{C}) = \infty$ a.e.

We shall be concerned with (finite state) shift automorphisms T of a double infinite product space $(X,\mathcal{B})$ with invariant probability m, $(X = \prod_{n=-\infty}^{\infty} \{1,2,\ldots k\})$. The states 1,2,...k define the state partition

$$\alpha = (A_1,\ldots A_k) \text{ where } A_i = \{x \in X : x_o = i\}.$$

If we have two such automorphisms T_1,T_2 then they are finitarily isomorphic if there is an isomorphism $\emptyset : X_1 \to X_2$ with $\emptyset T_1 = T_2\emptyset$ such that $\emptyset^{-1}\alpha_2$ consists of elements which are almost everywhere countable unions of cylinders of X_1 and such that $\emptyset\alpha_1$ consists of elements which are almost everywhere countable unions of cylinders of X_2.

Theorem [K.S.].

If T_1,T_2 are finite state Bernoulli automorphisms with equal entropy then they are finitarily isomorphic.

With the above notation let

$$\emptyset^{-1}A_i' = \cup C_n \quad \text{for some } A_i' \in \alpha_2 \text{ where the } C_n$$

are cylinders of the form $\{x \in X_1 : x_{-k} = i_k \ldots x_\ell = i_\ell\}$ with $k \geq 0$, $\ell \geq 0$ of minimal 'length'. For x belonging to such a cylinder we define the future code length of $\emptyset$ as $f(x) = \ell$.

If $a_n = m_1\{x : f(x) \geq n\}$ then the expected code length of $\emptyset$ is

$$\int f(x)dm_1 = \sum_{n=1}^{\infty} a_n.$$

It is not difficult to see that if α_1^- denotes the σ-algebra generated by $\bigcup_{n=0}^{\infty} T^{-n}\alpha_1$ then $d(\emptyset^{-1}\alpha_2, T_1^{\,n}\alpha_1^-) \leq 2a_{n+1}$ and hence, writing $\alpha = \alpha_1$, $\beta = \emptyset^{-1}\alpha_2$ and using the basic properties of d

$$d(\beta^-, T_1^{\,n}\alpha^-) \leq 2 \sum_{i=n+1}^{\infty} a_i .$$

§2. Information as a cocycle

In this section we use the functions $I_{T_1} = I(\alpha|T_1^{-1}\alpha^-)$ and $I_{T_2} = I(\alpha_2|T_2^{-1}\alpha_2^-) = I(\beta|T_1^{-1}\beta^-)\circ\emptyset^{-1}$ to ascertain the possibility of finite expected code and inverse code lengths. We retain the notation of §1 with the assumption that $\emptyset$ is a finitary isomorphism.

Lemma 2. If $\emptyset$ is a finitary isomorphism between T_1 and T_2, with finite expected future code length then $I(\beta^-|\alpha^-) < \infty$ on a set of positive measure.

Proof. Since $\emptyset$ has finite expected future code length $d(\beta^-, T_1^n\alpha^-) \leq 2\sum_{i=n+1}^{\infty} a_i < 2$ for n large enough. For such n, by lemma 1, we have $I(\beta^-|T_1^n\alpha^-) < \infty$ on a set of positive measure. However,

$$I(\beta^-|T_1^{n-1}\alpha^-) \leq I(\beta^- \vee T_1^n\alpha|T_1^{n-1}\alpha^-) = I(\beta^-|T_1^n\alpha^-) + I(T_1^n\alpha|T_1^{n-1}\alpha^-)$$

and since this latter is finite on a set of positive measure, we have by induction $I(\beta^-|\alpha^-) < \infty$ on a set of positive measure.

A finite valued function of the form $g\circ T_1 - g$ is called a T_1 coboundary. If two functions differ by a coboundary they are said to be cohomologous.

Theorem 1. Let T_1, T_2 be ergodic. If the finitary isomorphism $\emptyset$, and its inverse have finite expected code lengths then I_{T_1} and $I_{T_2} \circ \emptyset$ are cohomologous.

Proof. $I(\alpha^- \vee \beta^- | T_1^{-1}\alpha^-) = I(\alpha^- | T_1^{-1}\alpha^-) + I(\beta^- | \alpha^-)$

$$= I(\alpha^- \vee \beta^- \vee T^{-1}\beta^- | T_1^{-1}\alpha^-) = I(\alpha^- \vee \beta^- | T_1^{-1}\alpha^- \vee T_1^{-1}\beta^-) + I(\beta^- | \alpha^-) \circ T_1.$$

Ergodicity implies that all these quantities are finite a.e..

Hence $I(\alpha | T_1^{-1}\alpha^-)$ and $I(\alpha \vee \beta | T_1^{-1}\alpha^- \vee T_1^{-1}\beta^-)$ are cohomologous. A similar argument showsthat $I(\beta | T_1^{-1}\beta^-)$ and $I(\alpha \vee \beta | T_1^{-1}\alpha^- \vee T_1^{-1}\beta^-)$ are cohomologous.

§3. Invariants of the cocycle-coboundary equation.

We continue with the assumption that $\emptyset$ is a finitary isomorphism with finite expected code and inverse code lengths. By §2 we have

(3.1) $$I_{T_1} = I_{T_2} \circ \emptyset + g \circ T_1 - g$$

for some finite valued g. Our problem is how to exploit this relationship. One way, as in Bowen's paper [B] (although Bowen's context was slightly different) is to use the central limit theorem, if valid for either

$I^o_{T_1} = I_{T_1} - h(T_1)$ $I^o_{T_2} = I_{T_2} - h(T_2)$. The result which emerges is

$$\sigma^2(T_1) \equiv \lim_{n\to\infty} \frac{1}{n} \int |I^o_{T_1} + \ldots + I^o_{T_1} T_1^{\,n}|^2 \equiv \sigma^2(T_2).$$

This invariant is sufficient for Meschalkin's examples $(\frac{1}{4},\frac{1}{4},\frac{1}{4},\frac{1}{4})$, $(\frac{1}{2},\frac{1}{8},\frac{1}{8},\frac{1}{8},\frac{1}{8})$. In other words, no finitary isomorphism between these Bernoulli automorphisms has the property that both it and its inverse have finite expected code length.

If for some reason we also know that the functions in (3.1), including g, are L^2 functions then the fact the $\sigma^2(T_1)$ is an invariant of (3.1) can be proved directly. (Cf [F.P.].)

Another invariant of (3.1) is the group

$$\Lambda(T_1) = \{(a,b) \quad R \times R : \exp 2\pi i(a + ibI_{T_1}) = F \circ T_1/F$$

$$\text{for some measurable } F : X_1 \to \text{circle}.$$

This invariant is readily computable for Markov automorphisms. Since it can be proved that if $F \circ T_1/F$ is a function of n + 1 consecutive coordinates then F is a function of n coordinates (cf [P.2].) The Markov automorphisms $\begin{pmatrix} pq \\ pq \end{pmatrix}$, $\begin{pmatrix} pq \\ qp \end{pmatrix}$, $\begin{pmatrix} qp \\ pq \end{pmatrix}$ $(p \neq q)$ have identical σ^2 invariants but distinct Λ invariants. Again no two of these can be finitarily encoded with finite code and inverse code lengths.

References

[B] R. Bowen. Smooth partitions of Anosov diffeomorphisms are weak Bernoulli. Israel J. Math. 21(1975) 95-100.

[F.P.] R. Fellgett & W. Parry. Endomorphisms of a Lebesgue space II. Bull. L.M.S. 7(1975), 151-158.

[K.S.] M. Keane & M. Smorodinsky. Bernoulli schemes of the same entropy are finitely isomorphic. (To appear.)

[P.1.] W. Parry. Finitary isomorphisms with finite expected code lengths. (To appear.)

[P.2.] W. Parry. Endomorphisms of a Lebesgue space III. Israel J. Maths. 21(1975), 167-172.

THE LORENZ ATTRACTOR AND A RELATED POPULATION MODEL

William Parry

Introduction

A number of papers have emerged quite recently [G.], [W.], [R.], which model their investigations of "turbulence" on the Lorenz attractor [L.], (cf. also [R.T.]). The latter is obtained as the inverse limit flow $\hat{S}_t$ of the *Lorenz semi-flow* S_t whose phase-portrait looks like:

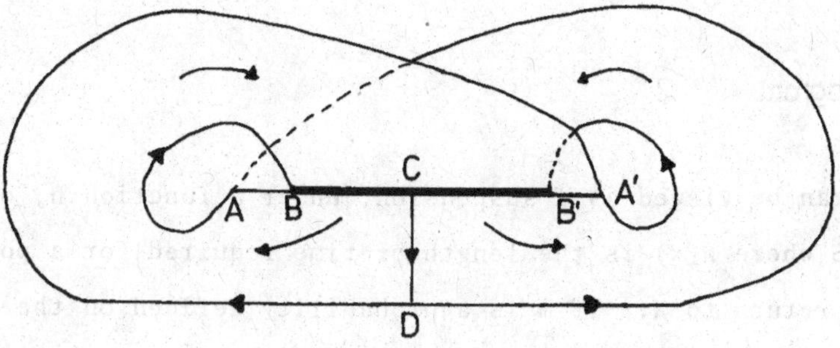

Diagram 1

We propose to investigate the analysis and geometry of such a turbulent semi-flow. When the above diagram is centrally symmetric the semi-flow S_t' obtained by factoring with respect to this Z_2 symmetry is closely related to a specific population growth model which we shall also comment upon.

Diagram 1 portrays a two-dimensional branched manifold M with branch set BB'. If $X = AA' = [0,1]$ and S denotes the Poincaré map of first return to X then S is well defined except at C. (C moves under S_t towards the stationary saddle point D.) The graph of S looks like:

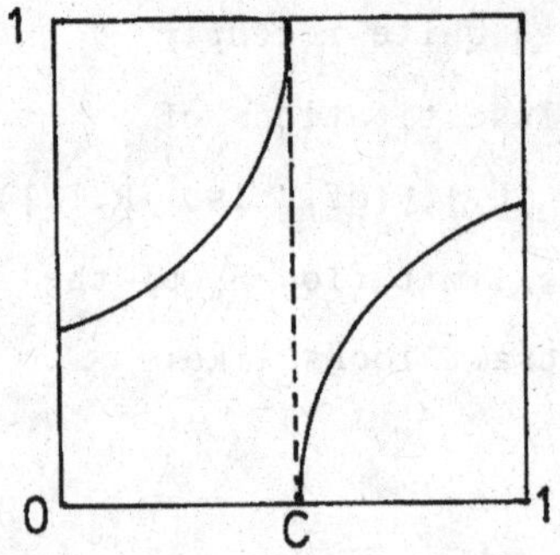

Diagram 2

Thus S_t can be viewed as a suspension, under a function h, of the map S where h(x) is the length of time required for a point $x \in X$ to return to X. If m is a probability defined on the Borel sets of X which is S invariant $(m(S^{-1}B) = (B)$ for all Borel sets B) then h together with m gives rise to an S_t invariant probability p and p then determines an $\hat{S}_t$ invariant probability $\hat{p}$. The spectral properties of $\hat{S}_t$ (with respect to $\hat{p}$) are determined easily from those of S_t (with respect to p). However, only when h is constant (say $h \equiv 1$), can one without extra effort deduce the spectral properties of S_t from those of S. When $h \equiv 1$ the representation of R^+ given by

$$U_t f = f \circ S_t \quad (f \in L^2(M,p))$$

is the representation induced (in Mackey's sense) by the

sub-semi-group $Z^+ \subset R^+$

$$U^n f = f \circ S^n \quad (f \in L^2(X,m)).$$

The spectral measures of U_t (defined on R) are obtained from the spectral measures of U (defined on the circle or [0,1)) by periodic continuation.

Now that the interelationships between $\hat{S}_t$, S_t and S have been roughly described we shall concentrate our attention on S with occasional references to S_t. Our purpose is to give a spectral and geometric description of the dynamics of S and a related population model.

Before proceeding, I would like to thank David Rand for posing the spectral problem to me and for informing me that some experimental data seemed to indicate a convergence to zero of correlation functions - a fact we shall establish for certain Poincaré-Lorenz maps S. (For others we shall establish the existence of eigenvalues with dyadic-rational roots of unity.) Special thanks are due to Bob Williams who pointed out to me, more than a year ago, the possible relevance of some of my early papers' to the Lorenz attractor. We shall assume that the reader is familiar with the notions and basic properties of entropy, weak-Bernoulli transformations, Lebesgue spectrum and strong-mixing. (cf. for example [S].)

§1. Linearization

Williams has shown [W], with a very simple argument, that if S is differentiable (except at C) and $S'(x) > \sqrt{2}$ for all $x \in X$ then S is *leo* i.e. S is locally, eventually, onto: for every interval $I \subset X$ there exists n such that $\bigcup_{i=0}^{n} S^n(I) = X$. This was precisely the property we used in [P1] to prove:

Proposition 1

If a piecewise continuous map S (with a finite number of turning points) of X onto itself is leo then S is topologically conjugate to a piecewise linear map T of X onto itself where the absolute value β of the slope of T is constant.

Thus from a qualitative point of view there is some justification in assuming that the Poincaré-Lorenz map T is piecewise-linear with constant absolute slope β. Hence T has a graph

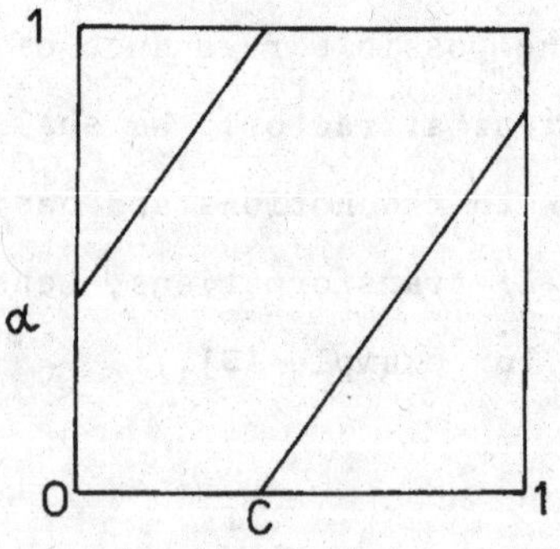

Diagram 3

With or without the justification of Proposition 1 we shall henceforth assume that T has the graph of diagram 3. This is the right moment, perhaps, to say something about C. As a Poincaré map T is not defined on C since C never returns to AA'. Bearing this in mind, we shall sometimes define $T(C) = 0$ and consider $X = [0,1)$. Hence T takes on the form $Tx = \beta x + \alpha \bmod 1$, with $\beta > 1$ and $0 \le \alpha < 1$. This will make no difference to our measure theoretic or spectral investigations since all our measures will be absolutely continuous and therefore will assign zero to countable sets. An invariant probability m absolutely continuous with respect to Lebesgue measure, for such transformations, was constructed in [P2]. (A problem concerning the possibility of m being a *signed* measure arose for a certain range of β ; however, this was cleared up in [H].)

<u>Proposition 2</u>. [P2], [H], [G], [P3].

If $Tx = \beta x + \alpha \bmod 1$ where $\beta > 1$ and $0 \le \alpha < 1$, then

$$m(E) = \int_E h(x)dx$$

is a T invariant probability where

$$h(x) = \frac{1}{\Delta} \sum_{n=0}^{\infty} \beta^{-n} \sum_{n=0}^{\infty} (\chi_{[0,T^n(1))}(x) - \chi_{[0,T^n(0))}(x)).$$

Δ is a normalising factor ($\int_0^1 h(x)dx = 1$) and $T(1) = \beta + \alpha \bmod 1$.

Wilkinson [W'] showed that with respect to the above measure m, T is weak-Bernoulli and therefore T has Lebesgue-spectrum in $L^2(X,m) - \mathbb{C}$ (the square integrable functions with zero integral,) when $\beta \geq 2$. Moreover m is equivalent to Lebesgue measure when $\beta \geq 2$ (i.e. $h(x) > 0$ a.e.). This result can be improved to some extent using a theorem of Bowen's [B]:

<u>Proposition 3</u>.

If T is a piecewise C^2 map of the unit interval onto itself with $T'(x) > \sqrt{2}$ with two interval pieces, then T is weak-Bernoulli with respect to a unique probability absolutely continuous to Lebesgue measure.

<u>Corollary</u>.

If $\beta > \sqrt{2}$ and $\beta + \alpha \leq 2$ $(\alpha \geq 0)$ then $Tx = \beta x + \alpha \bmod 1$ is weak-Bernoulli with respect to the probability as defined in Proposition 2.

<u>Proof</u>.

In this case the interval $[0,\frac{1-\alpha}{\beta})$ is mapped linearly to $[\alpha,1)$ and the interval $[\frac{1-\alpha}{\beta},1)$ is mapped linearly to $[0,\beta+\alpha-1)$ in a one-one fashion. Thus Bowen's conditions are satisfied.

§2. Centrally symmetric P.L. maps.

A P.L. map T (Poincaré-Lorenz or Piecewise-linear) $Tx = \beta x + \alpha \bmod 1$ is *centrally-symmetric* if $\beta + 2\alpha = 2$ ($\alpha = 1-\frac{\beta}{2}$), $1 < \beta \le 2$) so that T has a graph:

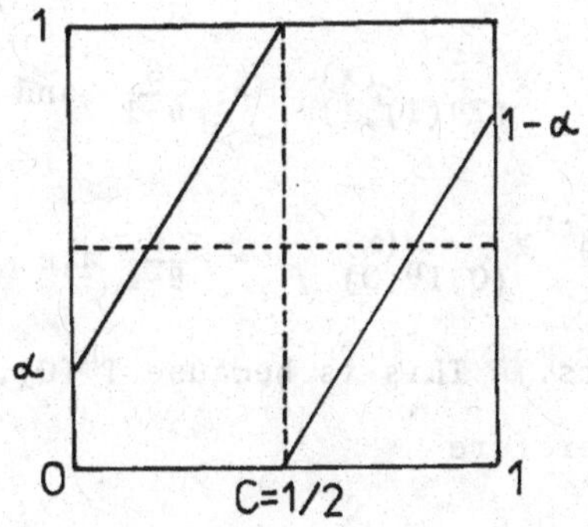

Diagram 4

In this case, we shall show that $h(x) > 0$ a.e. and therefore that m is equivalent to Lebesgue measure, when $\beta \ge 2^{\frac{1}{2}}$.

Proposition 4.

If $Tx = \beta x + 1 - \frac{\beta}{2} \bmod 1$ where $2^{\frac{1}{2}}$ 2 then $h(x) > 0$ a.e.

Proof. We shall leave the special case $\beta = 2^{\frac{1}{2}}$. It is particularly simple as it gives rise to an irreducible Markov chain with period 2.

Let $\beta > 2^{\frac{1}{2}}$ and

$$f(x) = \sum_{n=0}^{\infty} \beta^{-n} \chi_{[0,T^n(1))}(x)$$

$$g(x) = \sum_{n=0}^{\infty} \beta^{-n} \chi_{[0,T^n(0))}(x) \quad .$$

Then $h(x) = f(x) - g(x)$. Note that

$$f(x) + \sum_{n=0}^{\infty} \beta^{-n} \chi_{[T^n(1),1)}(x) = \frac{\beta}{\beta-1} \text{ and}$$

$$f(1-x) + \sum_{n=0}^{\infty} \beta^{-n} \chi_{(0,T^n(0)}(x) = \frac{\beta}{\beta-1} \text{ a.e. (in fact, except}$$

at countably many points.) This is because $T^n(0)$, $T^n(1)$ are symmetric about $\frac{1}{2}$. Therefore

$$f(1-x) + g(x) = \frac{\beta}{\beta-1} \quad \text{a.e. and}$$

$$h(x) = f(x) - g(x) = f(x) + f(1-x) - \frac{\beta}{\beta-1} \quad .$$

By the symmetry of h we need only consider $0 \le x \le \frac{1}{2}$. Using the fact that f is decreasing and the fact that $f \ge 1$ we have

$$\begin{aligned} h(x) = f(x) = f(1-x) - \frac{\beta}{\beta-1} &\ge f(\tfrac{1}{2}) + 1 - \frac{\beta}{\beta-1} \\ &\ge 1 + \frac{1}{\beta} + 1 - \frac{\beta}{\beta-1} \\ &= \frac{(\beta+1)}{\beta} - \frac{1}{\beta-1} \\ &= \frac{\beta^2-2}{\beta(\beta-1)} > 0. \end{aligned}$$

§3. The range $\beta < 2^{\frac{1}{2}}$

In the last section we analysed centrally symmetric P.L. maps $T = T_\beta$ when $\beta \geq 2^{\frac{1}{2}}$. In this range the invariant probability m is equivalent to Lebesgue measure. For $\beta > 2^{\frac{1}{2}}$. T is 'chaotic' in the sense that T has very strong mixing properties; in fact T is weak-Bernoulli. In the limiting case $\beta = 2^{\frac{1}{2}}$, T is a periodic Markov chain with period 2 so that T^2 is weak-Bernoulli on each of two periodic sets. The periodic sets, of course, provide an eigen-function with -1 as an eigenvalue. We proceed now to the range $\beta < 2^{\frac{1}{2}}$.

In this section we consider $Tx = \beta x + (1-\beta/2) \bmod 1$ when $\beta^2 < 2$. Strictly speaking we are interested in the Poincaré map T of the Lorenz semi-flow with central symmetry and with T piecewise linear (whose expansive constant β satisfies $\beta^2 < 2$) so that T is not defined at $\frac{1}{2}$.

Let $\alpha = 1-\beta/2$ and define $\gamma = T(1-\alpha)$ then
$0 < \gamma < \alpha < \frac{1}{2}$ i.e. $0 < \alpha + \beta^2/2 - 1 < 1 - \beta/2 < 1/2$
since $1 < \beta < 2^{\frac{1}{2}}$. The map T of the unit interval X to itself is pictured as follows:

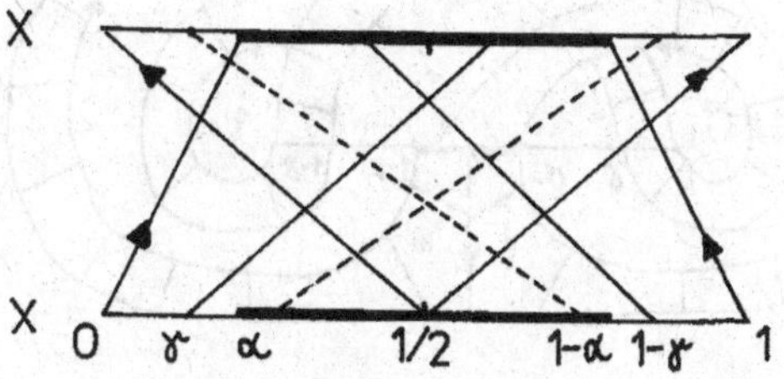

Diagram 5

Finally we note that $\frac{1}{2} < T(\gamma) < 1-\alpha$

i.e. $\frac{1}{2} < T(\beta(\beta-1)/2) < 1-\alpha \quad = \quad \beta/2$

since $1/2 < \beta^2(\beta-1)/2 + \alpha < 1-\alpha$

or $1 < \beta^2(\beta-1) + 2-\beta < \beta$. The last inequality is equivalent to $0 < (\beta-1)(\beta^2-1)$ and $\beta^2(\beta-1) < 2(\beta-1)$.

Let $I = (0,\gamma)$, $I' = (1-\gamma,1)$ and $J = (\alpha,1-\alpha) - \{\frac{1}{2}\}$. Then we have proved that J splits into symmetric associates $(\alpha,\frac{1}{2})$ and $(\frac{1}{2},1-\alpha)$ which are mapped one-one onto I', I respectively, by T. Moreover T maps each of I, I' one-one into J (neglecting the point $\frac{1}{2}$.)

Note: *We shall always discard points which are ultimately mapped to ½.*

Thus TJ is disjoint from J and $T^2J = J$. Moreover T^2 restricted to J is a centrally symmetric P.L. map of J onto itself with expansive constant β^2.

This statement is perhaps best summarised by the following diagram, where we have drawn lines transverse to the flow:

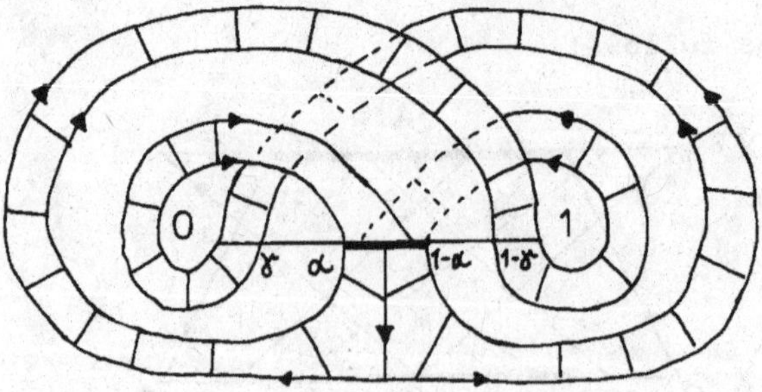

Diagram 6

Now let $T\varepsilon = 1-\varepsilon$ where $0 < \varepsilon < \frac{1}{2}$ then $\varepsilon = \dfrac{\beta}{2(\beta+1)}$ and obviously $T(1-\varepsilon) = \varepsilon$ so that $T^2(\varepsilon) = \varepsilon$, $T^2(1-\varepsilon) = 1-\varepsilon$. It is not difficult to see that ε is the only point with period 2 satisfying $0 < \varepsilon < \frac{1}{2}$.

Moreover $\gamma < \varepsilon < \alpha$ i.e. $\dfrac{\beta(\beta-1)}{2} < \dfrac{\beta}{2(\beta+1)} < 1 - \dfrac{\beta}{2}$, again because $\beta^2 < 2$. (These three-points coincide when $\beta^2 = 2$.)

Using the periodic point ε and the expansiveness of T we can see:

Orbits in the unshaded region not passing through ε, $1-\varepsilon$ *converge in an alternating fashion toward the shaded region and are captured by this region in a finite amount of time.*

We have described the geometry of the unshaded dissipative region. This has implications for any T invariant probability and in particular for m.

m must be supported by the conservative part of X and thus by the shaded part of X i.e. the intersection of the shaded area with X.

We have now reduced the analysis and geometry of T (or of T_t) to the analysis and geometry of T^2 restricted to $J = (\alpha, 1-\alpha)$. As we have already remarked T^2 is a centrally symmetric P.L. mapped which, when scaled up to the unit interval, has the form $T^2(x) = \beta^2 x + (1-\frac{\beta^2}{2}) \bmod 1$.

We can now analyse T^2 (on J) in the same way. Again we produce intervals I_1, I_1', J_1 (contained in J) with a dissipative complement and two periodic points of period 2. This procedure can be repeated until $\beta^{2^n} > 2^{\frac{1}{2}}$.
We have said enough to establish:

<u>Theorem</u>

Let $2^{\frac{1}{2^{n+1}}} < \beta < 2^{\frac{1}{2^n}}$ and let $T = T_\beta$ be the centrally symmetric P.L. map associated with the Lorenz semi-flow.
Then there exist disjoint sub-intervals I_k, I_k', J_k of X with J_k centrally symmetric and I_k' symmetrically associated with I_k and $J_k \supset I_{k+1} \cup I_{k+1} \cup J_{k+1}$ $k = 0, =, \ldots, n-1$ $(J_0 = X,\ I_0 = I_0' = \emptyset)$ such that T^{2^k} maps J_k onto itself $(k=0,=,\ldots,n)$ as described above for the case $k = 0$.
Thus there are exactly 2^k points of *least* period 2^k (in the range $k < n$).
The two points in J_k of period 2 for $T^{2^{k+1}}$ act as repellors and I_{k+1}, I_{k+1}', J_{k+1} are attractors.
T^{2^n} is weak-Bernoulli on J_n and on each of $T^i J_n$, $i = 1, \ldots 2^n-1$ as well,
Thus the spectral properties of T are described completely by the eigen-functions (with eigenvalues 2^nth roots of unity) provided by the period set J_n and its 2^n iterates and by the Lebesgue spectrum of T^{2n} on J_n.

The measure of Proposition 2 is equivalent to Lebesgue measure on J_n and its 2^n-iterates and is zero elsewhere. In particular if $f \in L^2(X,m)$ is concentrated on J_n with $\int f dm = 0$ then

$$\int f T^N \bar{f}\, dm = \int_0^1 e^{2\pi i t N} k(t) dt \to 0$$

where k is Lebesgue integrable. (This integral is actually *equal* to zero when N is not a multiple of 2^n.)

It is clear from this remark that the asymptotic behaviour of all correlations with respect to m, or with respect to Lebesgue measure, can easily be described using our Theorem and the fact that the weak-Bernoulli property implies a Lebesgue spectrum for functions $f \in L^2(X,m) \ominus \mathbb{C}$.

§3. The naturalness of the measure m.

If S is a Poincare-Lorenz map of a centrally symmetric Lorenz semi-flow and if S is piecewise C^2 and $S'(x) > \sqrt{2}$ for all $x \in X$ then by Proposition 1 S is topologically conjugate to a centrally symmetric P.L. map T ($\emptyset S = T\emptyset$, $\emptyset$ a homeomorphism.) Hence $m\emptyset$ is S invariant. However, there is no guarantee that $m\emptyset$ is absolutely continuous with respect to Lebesgue measure. Nevertheless a result in [L.Y.](cf. also [K]) provides an S invariant probability p which is absolutely continuous with respect to Lebesgue measure. S is weak-Bernoulli with respect to p. Thus S preserves and is weak-Bernoulli with respect to $m\emptyset$ and p. $m\emptyset$ and p may well be mutually singular. p clearly has the advantage in "naturalness" in one respect - its absolutely continuity. However, from another point of view, $m\emptyset$ is perhaps the most natural. Without going into details the m entropy of T is log β where β is the expansive constant of T and therefore the $m\emptyset$ entropy of S is log β . It is not difficult to show that the topological entropy of T and therefore of S is log β which is known to dominate all measure theoretic entropies. Hence $m\emptyset$ is a 'maximal' measure for S.

Conjecture: S has only one 'maximal' measure.

If this is true, we have an extra justification for studying P.L. maps.

§4. A population model

The map of $X = [0,1]$ into itself given by $f = f_b$, $f_b(x) = 4bx(1-x)$ $(0 \le b \le 1)$ has received a great deal of attention just recently in connection with population fluctuation. (cf. [H.S] for a derivation of this map from a differential equation for population dynamics, and [S.W.] for a recent study, together with its bibliography.)

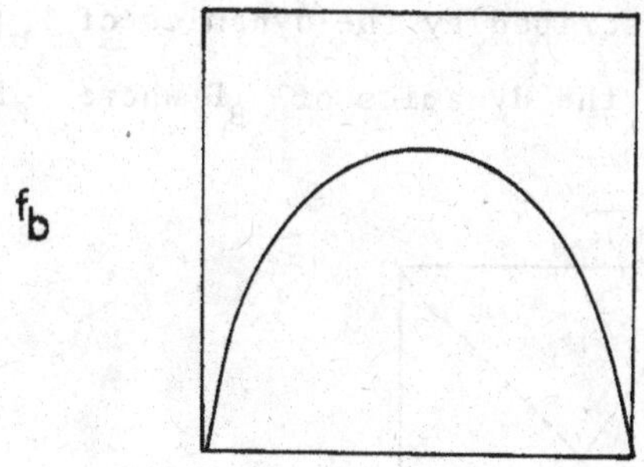

Diagram 7

We propose (with no apologies or justification) to study the piecewise linear analogue F_b of f_b.

$$F_b(x) = b(1-|1-2x|) \qquad (0 \le b \le 1)$$

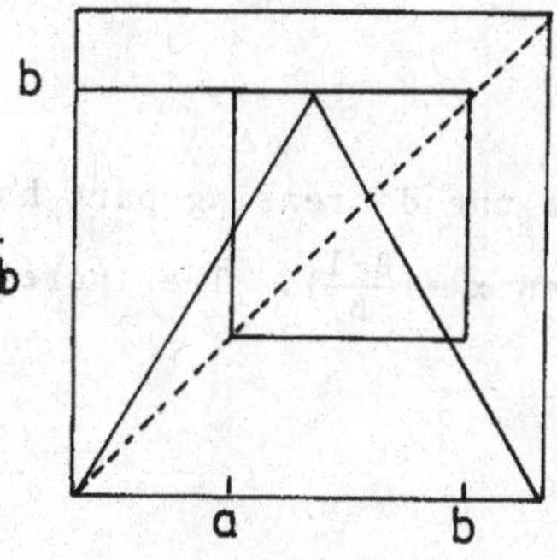

Diagram 8

The dynamics of F_b is not very interesting for $0 \le b < \frac{1}{2}$ for inthis case $F_b^n(x) \to 0$ for all x. If $b = \frac{1}{2}$, $F_b(x) = x$ when $0 \le x \le \frac{1}{2}$, and $F_b(x) = 1-x$ so that there is no movement after one iteration. We suppose then that $\frac{1}{2} < b \le 1$.

Let $a = F_b(b)$, then from diagram 8, it is clear that $(b,1]$ gets mapped into $[0,a)$ and all points $x \in [0,a)$ are attracted by iteration toward the interval $[a,b]$.

Thus the dynamics of F_b are described by the dynamics of $F_b|[a,b]$. Scaling up our problem becomes the dynamics of $_{\beta}F$ where $_{\beta}F$ has the graph:

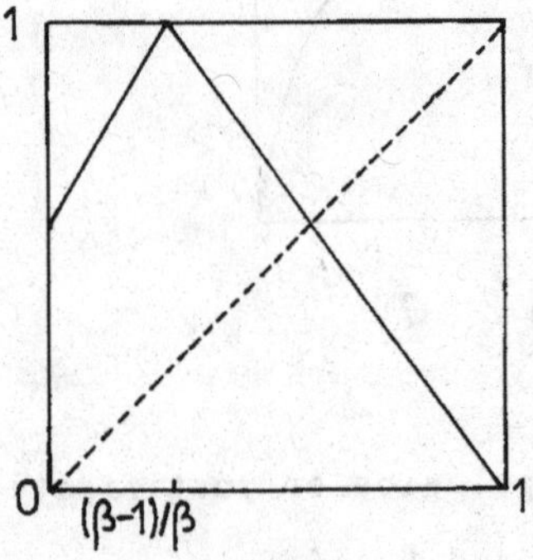

Diagram 9

F_b has slope $1 < \beta \le 2$ so that the decreasing part has equation $y = -\beta x + \beta$. Hence $y = 1$ when $x = (\frac{\beta-1}{\beta})$. The increasing part has equation $y = \beta x + 2-\beta$.

The map ${}_\beta F$ whose graph is diagram 9 can also be obtained as follows. Let T_β be the centrally symmetric P.L. map

$T_\beta x = \beta x + 1 - \frac{\beta}{2}$.

Since T_β is centrally symmetric the obvious Z_2 action $x \to 1-x$ commutes with T_β to produce a map of the interval $[0,\frac12]$ with graph

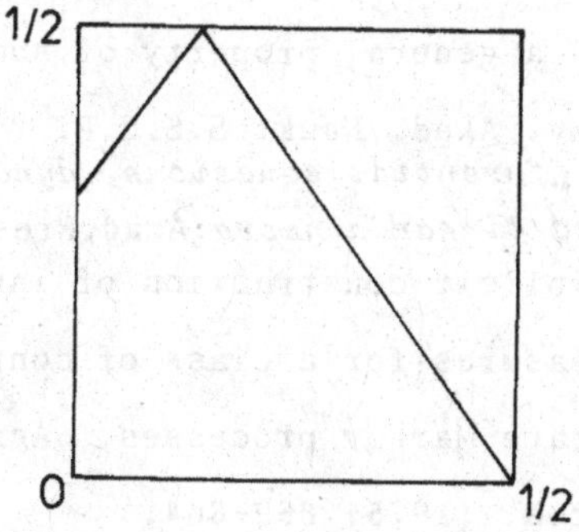

Diagram 10

This is the map ${}_\beta F$ except, again, for scale.

Thus the geometry and analysis of the P.L. map T_β gives information about the attractor of our population model. In particular when $2^{\frac12} < 2b \le 2$ (i.e. $\frac{1}{\sqrt{2}} < b \le 1$) the attractor is a region of chaos. We have provided the T invariant probability absolutely continuous with respect to Lebesgue measure. It is a symmetric probability. If we restrict this to $[0,\frac12]$, it provides a ${}_\beta F$ invariant measure assigning measure $\frac12$ to $[0,\frac12]$, so that doubling this measure will give the appropriate invariant probability for ${}_\beta F$.

REFERENCES

[B] R. Bowen — Bernoulli maps of the interval (To appear.)

[G] J. Guckenheimer — A strange, strange attractor. (To appear.)

[G'] A. O. Gelfond — On a general property of number systems, Izv. Akad. Nauk. S.S.S.R. 23 (1959) 809-814

[H.S.] M.Hirsch and S. Smale — *Differential equations, dynamical systems and linear algebra.* Acad.Press 1974 New York

[H] S. Halfin — Explicit construction of invariant measures for a class of continuous state Markov processes. Annals of Prob. (1975) 859-864.

[K] Z.S. Kowalski — Invariant measures for piecewise monotonic transformations. Springer Lecture Notes (472).

[L] E.N. Lorenz — Deterministic nonperiodic flow. Journal of the Atmos. Sc. 20 (1963) 130-141.

[L.Y.] A. Lasota and Y. York. — On the existence of invariant measures for piecewise monotonic transformations. Trans. Amer. Math. Soc. (1973) 481-488.

[Pl] W. Parry — Symbolic dynamics and transformations of the unit interval. Trans. Amer. Math. Soc. 122 (1966) 368-378.

[P2] W. Parry — Representations for real numbers. Acta. Math. Acad. Sci. Hung. (1964), 95-105.

[P3] W. Parry — On the β-expansions of real numbers. Acta. Math. Acad. Sci. Hung. (1960) 401-416.

[R] D. Rand — Kneading for the Lorenz attractor. Preprint, University of Warwick.

[R.T.] D. Ruelle and F. Takens — On the nature of turbulence. Commun. Math. Phys. 20 (1971) 167-192.

[S] P. Shields — The theory of Bernoulli shifts. Chicago Lectures in Math. 1973

[S.W.] S.Smale and R.F. Williams — The qualitative analysis of a difference equation of population growth. (To appear.)

[W.] R.F. Williams — The structure of Lorenz attractors. (To appear).

[W'.] K.M.Wilkinson — Ergodic properties of a class of piecewise linear transformations. Z. Wahrscheinlichkeitstheorie verw. Gebiete. 31. (1975), 303-328.

William Parry
Mathematical Institute
University of Warwick
Coventry, Warwick
CV4 7AL
U.K.

UNIQUE ERGODICITY AND RELATED PROBLEMS

Klaus Schmidt

1. Introduction

Let T be a Borel automorphism of a standard Borel space $(X,\mathcal{S})$. Throughout this note the word 'measure' will refer to a σ-finite Borel measure on $(X,\mathcal{S})$, even when this is not explicitly stated. A measure μ is said to be quasi-invariant (under T) if, for every $B \in \mathcal{S}$, $\mu(TB) = 0$ if and only if $\mu(B) = 0$. We shall consider the following two general problems in this connection:

(A) Does there exist a quasi-invariant measure μ on $(X,\mathcal{S})$ which has certain specified properties (e.g. invariance, nonatomicity and ergodicity, a specific type in the sense of weak equivalence, or a specified Radon-Nikodym derivative under T)?

(B) If such a measure μ exists, can one assume without loss of generality that μ is the only measure on $(X,\mathcal{S})$ with the specified properties? More precisely, does there exist a T-invariant Borel set $B \subset X$ such that $\mu(X\setminus B) = 0$, and $\nu(B) = 0$ for every measure ν on $(X,\mathcal{S})$ which is singular with respect to μ, but which otherwise shares its properties? A typical example is the classical result on unique ergodicit in its simplest form: If μ is an invariant ergodic probability measure on $(X,\mathcal{S})$, there exists a T-invariant Borel set $B \subset X$ with $\mu(B) = 1$ and with $\nu(B) = 0$ for every invariant and ergodic probability measure $\mu \neq \nu$ on $(X,\mathcal{S})$ (cf. [8] for a much stronger result).

This note is organized as follows: Section 2 deals with the problem under what conditions a countable group of Borel automorphisms of (X,S) has nonatomic, quasi-invariant, ergodic measures. Section 3 shows that the existence of one nonatomic, quasi-invariant and ergodic measure for a Borel automorphism T of (X,S) implies the existence of a wide variety of such measures, showing at the same time that even under fairly restrictive conditions one may expect a positive answer to (A), but a negative answer to (B). Section 4 deals with question (B) and shows that the classical unique ergodicity assertion stated there is, in fact, a special case of a more general result, according to which one can achieve unique ergodicity by specifying the Radon-Nikodym derivative of a quasi-invariant probability measure.

Most of the results in this note have appeared elsewhere, and I have therefore omitted their proofs. The article is meant as a brief survey of some recent attempts to throw light on the two problems (A) and (B). Some of the results mentioned extend to actions of countable groups of Borel automorphisms other than integer actions, and I will comment on this whenever it seems appropriate.

2. The existence of nonatomic ergodic measures

Let G be a countable group with identity e, which acts measurably on a standard Borel space (X,S). This means that there exists a map $(g,x) \to gx$ from $G \times X$ to X with $ex = x$, $g_1(g_2x) = (g_1g_2)x$ for every $x \in X$, $g_1, g_2 \in G$, and such that $x \to gx$ is a Borel automorphism of X for every $g \in G$. If X is a polish space and $x \to gx$ is a homeomorphism of X for every $g \in G$, we shall say that G acts continuously. In this section we consider the problem under which conditions there exists a nonatomic, quasi-invariant, ergodic probability measure for G on X. The following result is an immediate corollary to Theormes 2.6 and 2.9 in [3]:

Theorem 2.1. Let G be a countable group which acts continuously on a polish space X. The following conditions are equivalent:

(1) There exists a nonatomic probability measure μ on X which is quasi-invariant and ergodic under G;

(2) there exists no Borel cross section of the G-orbits in X (i.e. there exists no Borel set C in X which intersects every G-orbit in exactly one point).

Turning now to measurable actions, we shall establish a weaker, but otherwise analogous, result by using a completely different method to the one employed in [3]. The technique of the following proof is basically taken from [12, p.25].

Theorem 2.2. Let G be a countable group which acts measurably on a standard Borel space $(X,\mathcal{S})$. The following conditions are equivalent:

(1) There exists a nonatomic probability measure μ on X which is quasi-invariant and ergodic under G;

(2) there exists no universally measurable cross section of the G-orbits in X (i.e. there exists no subset $C \subset X$ which intersects every G-orbit in exactly one point, and which is measurable for every $\nu \in P(X)$, the set of all probability measures on $(X,\mathcal{S})$).

Proof: *For this proof we shall assume the continuum hypothesis.* Assume first that there exists no nonatomic probability measure on $(X,\mathcal{S})$ which is quasi-invariant and ergodic under G. Using standard decomposition theory, one can show that this implies the following (cf. [14, Chapter 6]): For every $\nu \in P(X)$ which is quasi-invariant under G, there exists a Borel set $B_\nu \subset X$ such that

$B_\nu \cap gB_\nu \cap \{x:gx \neq x\} = \emptyset$ for every $g \in G$ (i.e. B_ν intersects every G-orbit in at most one point), (2.1)

and

$$\nu(X\backslash GB_\nu) = 0, \tag{2.2}$$

where $GB_\nu = \bigcup_{g\in G} gB_\nu$. We choose and fix an enumeration $(g_1, g_2, \ldots)$ of G, define, for every $\mu \in P(X)$, a quasi-invariant probability measure $\tilde{\mu} \in P(X)$ by $\tilde{\mu}(B) = \sum_{k=1}^{\infty} 2^{-k} \mu(g_k B)$, $B \in S$, and choose a Borel set $B_{\tilde{\mu}}$ satisfying (2.1) and (2.2) with $\tilde{\mu}$ replacing ν. Using the continuum hypothesis we can choose a well-ordering $\prec$ of P(X) for which $\{\nu \in P(X): \nu < \mu\}$ is countable for every $\mu \in P(X)$. By induction one can prove the existence of a family of Borel sets C_μ, $\mu \in P(X)$, with the following properties:

(1) $C_\mu \cap gC_\mu \cap \{x: gx \neq x\} = \emptyset$ for every $g \in G$,

(2) $\tilde{\nu}(X\backslash GC_\mu) = \nu(X\backslash GC_\mu) = 0$ for every $\nu \leq \mu$,

(3) $C_\nu \subset C_\mu$ whenever $\nu < \mu$.

To give an explicit construction of such a family, let μ_1 be the first element in P(X), and put $C_{\mu_1} = B_{\tilde{\mu}_1}$. Assuming now that C_ν has been chosen for all $\nu < \mu$, put $C_\mu = \bigcup_{\nu<\mu} C_\nu \cup (B_{\tilde{\mu}} \setminus G(\bigcup_{\nu<\mu} C_\nu))$. C_μ is a Borel set, since μ has only countably many predecessors, and it is clear that it will again satisfy conditions (1) - (3). Having constructed C_μ for every $\mu \in P(X)$, let $C = \bigcup_{\mu\in P(X)} C_\mu$. C will again intersect every G-orbit in at most one point. Furthermore we have, for every $\mu \in P(X)$, $C\backslash C_\mu \subset X\backslash GC_\mu$, and $\mu(X\backslash GC_\mu) = 0$, so that C differs from the Borel set C_μ by a μ-null set. In other words, C is universally measurable. Finally, if μ_x is the probability measure concentrated in the point $x \in X$, we have $\tilde{\mu}_x(X\backslash GC) \leq \tilde{\mu}_x(X\backslash GC_{\mu_x}) = 0$, so that C intersects the orbit through x. Hence C is a universally measurable cross section of the G-orbits.

Conversely, if there exists a universally measurable cross section for G, it is an easy exercise to show that every quasi-invariant, ergodic, σ-finite measure for G must be atomic, so that the theorem is proved completely.

Remark 2.3. I do not know whether one can replace universal measurability of the cross section C by the condition that C be Borel.

Having established a necessary and sufficient condition for a Borel automorphism T of a standard Borel space $(X,\mathcal{S})$ to have at least one nonatomic, quasi-invariant, ergodic measure it is natural to ask if there are other, essentially different, measures with these properties. The next section will show that there must exist nonatomic, quasi-invariant and ergodic measures satisfying a wide variety of conditions. I.P. Kornfel'd [6] uses a different starting point to arrive at similar conclusions. He shows that a homeomorphism of a polish space has a nonatomic, quasi-invariant, ergodic measure if and only if it has a (suitably defined) recurrent point, and he then proceeds to show that in this case there must exist many such measures.

3. Quasi-invariant ergodic measures with specified properties

Let T be a Borel automorphism of a standard Borel space $(X,\mathcal{S})$ which admits a nonatomic, quasi-invariant, ergodic probability measure μ. If T' is another nonsingular ergodic automorphism of a standard probability space $(X',\mathcal{S}',\mu')$, we call T and T' *weakly equivalent* if there exists a nonsingular bijection $\phi: x \to x'$ such that, for μ-a.e. $x \in X$, $\phi(\{T^n x : n \in Z\}) = \{T'^n \phi(x) : n \in Z\}$. The following theorem, due to W. Krieger [10], gives some idea of the variety of ergodic measures for T.

Theorem 3.1. Let T' be any nonsingular ergodic automorphism of a standard nonatomic probability space $(X',\mathcal{S}',\mu')$. Then there exists a nonatomic, quasi-invariant, ergodic probability measure ν on (X,S) such that T, acting on (X,S,ν), is weakly equivalent to T'.

Another way of imposing conditions on a nonatomic, quasi-invariant ergodic measure ν for T on (X,S) is to specify the Radon-Nikodym derivative $\rho_\nu(x) = \log \frac{d\nu T}{d\nu}(x)$. We need two definitions: A Borel function $f:X \to R^n$ $(n \geq 1)$ is said to be a *coboundary* for T on (X,S,ν) if there exists a Borel function $h:X \to R^n$ such that $f(x) = h(Tx)-h(x)$ for ν-a.e. $x \in X$. f is called *recurrent* for T on (X,S,ν) if, for every Borel set $B \subset X$ with $\nu(B) > 0$, and for every $\varepsilon > 0$, $\nu(\bigcup_{m\geq 1} (B \cap T^{-m}B \cap \{x: \| \sum_{k=0}^{m-1} f(T^k x)\| < \varepsilon\})) > 0$. Here $\|.\|$ stands for the usual norm on R^n. For a detailed discussion of these defintions refer to [14]. The following lemma is due to W. Krieger ([7], cf. also [14, Theorem 4.2]):

Lemma 3.2. Let ν be a nonatomic, quasi-invariant and ergodic measure for T on (X,S), and let $\rho_\nu:X \to R$ be a Borel function with

$$\rho_\nu(x) = \log \frac{d\nu T}{d\nu}(x) \quad \nu\text{-a.e.} \tag{3.1}$$

Then ρ_ν is recurrent for T on (X,S,ν).

The second main result that we quote in this section is again an indication of the enormous diversity of ergodic measures for T.

Theorem 3.3. Let μ be a nonatomic, quasi-invariant and ergodic measure for T on (X,S), let $\rho_\mu:X \to R$ be given by (3.1), and let $f:X \to R$ be a Borel map such that the pair $(\rho_\mu,f):X \to R^2$ is recurrent for T on (X,S,μ). Then there exists an uncountable index set J and a family $\{\nu_\alpha:\alpha \in J\}$ of infinite, nonatomic, in variant and ergodic measures for T on (X,S) with the following properties:

(1) ν_α and ν_β are mutually singular whenever $\alpha \neq \beta$,

(2) for every $\alpha \in J$, (ρ_μ,f) is a coboundary for T on (X,S,ν_α),

(3) if $B \in S$ satisfies $\nu_\alpha(B) = 0$ for every $\alpha \in J$, then $\mu(B) = 0$.

Proof: This is Theorem 10.5 in [14], applied to the function (ρ_μ, f), and Corollary 10.6 of [14] now reads as follows:

Corollary 3.4. There exist uncountably many mutually singular nonatomic, quasi-invariant, ergodic measures $\{\sigma_\alpha : \alpha \in J\}$ for T on (X,S) such that conditions (1) and (3) in Theorem 3.3 are satisfied, and

(2') for every $\alpha \in J$, $\log \frac{d\sigma_\alpha T}{d\sigma_\alpha}(x) = \rho_\mu(x)$ σ_α-a.e.

Proof: Putting $f = 0$ and applying Lemma 3.2, we can use Theorem 3.3 to obtain a family $\{\nu_\alpha : \alpha \in J\}$ with the properties stated there. In particular there must exist, for every $\alpha \in J$, a Borel map $h_\alpha : X \to R$ with $\rho_\mu(x) = h_\alpha(Tx) - h_\alpha(x)$ for ν_α-a.e. $x \in X$. We put $d\sigma_\alpha(x) = e^{h_\alpha(x)} . d\nu_\alpha(x)$, and the corollary is proved.

Similarly one proves

Corollary 3.6. Let $f: X \to R$ satisfy the conditions of Theorem 3.3. Then there exists an uncountable family $\{\sigma_\alpha : \alpha \in J\}$ of nonatomic, quasi-invariant, ergodic measures for T on (X,S) which satisfy conditions (1) and (3) in Theorem 3.3, and for which

(2") $\log \frac{d\sigma_\alpha T}{d\sigma_\alpha}(x) = f(x)$ σ_α-a.e., for every $\alpha \in J$.

Remark 3.7. If μ is a nonatomic, ergodic, T-invariant probability measure, and if $f: X \to R$ is a Borel map with $\int f \, d\mu = 0$, then the requirements of Theorem 3.3 (or Corollary 3.6) are fulfilled (cf. [1] or [14, Theorem 11.4].

The results of this section show - among other things - that one can never single out a unique measure on a standard Borel space by specifying either its type in the sense of weak equivalence, or its

Radon-Nikodym derivative under T. Various results in this direction were achieved in [4], [5], [6],[7], [10], [11], [13]. All these methods depend deeply on the fact that there is a single transformation acting on the space (more precisely: that the orbit structure is hyperfinite). A recent result which goes beyond the hyperfinite case and which seems of a potentially much more general nature is due to S.G. Dani and M. Keane [2]. They show that, for example, the group SL(2,Z), acting on the 2-torus, has uncountably many nonatomic, infinite, invariant, ergodic measures.

4. Unique ergodicity for quasi-invariant probability measures

Having seen that unique ergodicity is not a meaningful concept even for infinite invariant measures on a standard Borel space, we now turn to the second problem mentioned in the introduction. As it turns out, it is always possible to single out a unique *probability* measure with specified Radon-Nikodym derivative. To make this clear, let $f:X \to R$ be a Borel map and let $P_f(X)$ denote the set of all $\nu \in P(X)$ which are quasi-invariant under T and which satisfy

$$\log \frac{d\nu T}{d\nu}(x) = f(x) \quad \nu\text{-a.e.} \tag{4.1}$$

For every Borel function f, $P_f(X)$ is either empty or a convex Borel subset of P(X).P(X) is endowed with its natural Borel structure which makes the map $\nu \to \int h\, d\nu$ measurable for every bounded Borel function h:X → R). The problem whether $P_f(X)$ is nonempty is completely analogous to the problem whether T admits an invariant probability measure - i.e. whether $P_0(X)$ is nonempty, where 0 denotes the constant function 0 on X. The following result is due to H. Furstenberg (oral communication).

<u>Theorem 4.1</u>. Let T be a homeomorphism of a compact metric space X, and let $f:X \to R$ be a continuous function. Then there exists a constant c such that $P_{f+c}(X) \neq \emptyset$. If T is uniquely ergodic with a unique

invariant probability measure μ, then $P_f(X) \neq \emptyset$ if and only if $\int f \, d\mu = 0$.

Proof: In its usual topology $P(X)$ is a compact, convex, metric space, and the map $u: P(X) \to P(X)$. given by

$$d(u(\nu))(x) = e^{-f(x)} . (\int e^{-f} \, d\nu T)^{-1} . d\nu T(x),$$

is a homeomorphism. By the Schauder fixed point theorem there exists a fixed point ν_o for this map. We put $c = \log (\int e^{-f} \, d\nu_o T)$ and observe that $\nu_o \in P_{f+c}(X)$. The second assertion follows from Lemma 3.2: due to the unique ergodicity of T, a continuous function f is recurrent for *any* quasi-invariant measure if and only if $\int f \, d\mu = 0$. Hence the constant c in the first part must be equal to zero whenever $\int f \, d\mu = 0$. The proof is complete.

In general we cannot expect $P_f(X)$ to be nonempty. This is emphasized by the next theorem [15, Theorem 1.3].

Theorem 4.2. Let T be a nonsingular ergodic automorphism of a standard probability space (X,S,μ), and let $f: X \to R$ be a Borel map with the following property: there exists no μ-integrable Borel function $h: X \to R^+$ (the positive real numbers) with $\frac{d\mu T}{d\mu}(x) = e^{f(x)} . h(x) . h(Tx)^{-1}$ for μ-a.e. $x \in X$. Then there exists a T-invariant Borel set $B \subset X$ with $\mu(B) = 1$ and with $\nu(B) = 0$ for every $\nu \in P_f(X)$.

Corollary 4.3. Let T be an ergodic measure preserving automorphism of a standard measure space (X,S,μ) and let $f: X \to R$ be a Borel function which is not a coboundary for T on (X,S,μ). Then there exists a T-invariant Borel set B in X with $\mu(X \setminus B) = 0$ and with $\mu(B) = 0$ for every $\nu \in P_f(X)$.

If we consider measure preserving ergodic automorphisms only up to conjugacy (as one usually does), we may assume $P_f(X)$ to be empty whenever f is not a coboundary. An analogous statement holds for non-measure-preserving automorphisms. In particular we have:

Corollary 4.4. Let T be a nonsingular ergodic automorphism of a standard measure space (X,S,μ). If μ is not equivalent to a T-invariant probability measure, there exists a T-invariant Borel set B in X with $\mu(X\setminus B) = 0$ and with $\nu(B) = 0$ for every T-invariant probability measure ν on X.

Turning finally to the problem of singling out a particular probability measure by its Radon-Nikodym derivative, one can prove the following result [15, Theorem 1.2]:

Theorem 4.5. Let T be a Borel automorphism of a standard Borel space (X,S) and let $f:X \to R$ be a Borel map with $P_f(X) \neq \emptyset$. Then there exists, for every ergodic $\mu \in P_f(X)$, a T-invariant Borel set $B_\mu \subset X$ with the following properties:

(1) $\mu(B_\mu) = 1$,

(2) if $\nu \in P_f(X)$ is ergodic and $\nu \neq \mu$, then $\nu(B_\mu) = 0$.

Theorem 4.5 shows that, for every ergodic $\mu \in P_f(X)$, we may assume T to be *uniquely ergodic within the class* $P_f(X)$, at least when restricted to a suitable T-invariant subset of full measure. It is interesting to note that all the results of this section have straightforward analoga for actions of arbitrary countable groups, and have nothing to do with hyperfiniteness.

REFERENCES

1. Atkinson, G,: Recurrence of cocycles and random walks. J. London Math. Soc . (2), 13 (1976), 486-488.

2. Dani, S.G., and Keane, M.: Ergodic invariant measures for actions of SL(2,Z). Preprint (1978).

3. Effros, E.G.: Transformation groups and C*-algebras. Ann. of Math. 81 (1965), 38-55.

4. Katznelson, Y., and Weiss, B.: The construction of quasi-invariant measures. Israel J. Math. 12 (1972), 1-4.

5. Keane, M.: Sur les mesures quasi-ergodiques des translations irrationelles. C.R.Acad.Sci.Paris 272 (1971),54-55.

6. Kornfel'd, I.P.: Quasi-invariant measures for topological dynamical systems (Russian). Izv.Akad.Nauk.SSSR Ser.Mat. 38 (1974), 1305 - 1323.

7. Krieger, W.: On quasi-invariant measures in uniquely ergodic systems. Inventiones Math. 14 (1971), 184-196.

8. ———— : On unique ergodicity. Proc. Sixth Berkeley Symp. on Math. Stat. and Probability. Berkeley-Los Angeles: University of California Press, 1972, 327-346.

9. ———— : On ergodic flows and the isomorphism of factors. Math. Ann. 223 (1976), 19-70.

10. ———— : On Borel automorphisms and their quasi-invariant measures. Math. Z. 151 (1976), 19-24.

11. Mandrekar, V., and Nadkarni, M.: On ergodic quasi-invariant measures on the circle group. J.Functional Analysis 3 (1968), 157-163.

12. Oxtoby, J.C.: Measure and Category. New York-Heidelberg-Berlin: Springer, 1971.

13. Schmidt, K.: Infinite invariant measures on the circle. Symposia Math. XXI (1977), 37-43.

14. ———— : Cocylces of ergodic transformation groups. Macmillan (India) 1977.

15. ———— : Unique ergodicity for quasi-invariant measures. Preprint (1978).

Klaus Schmidt
Mathematical Institute
University of Warwick
Coventry, Warwick
CV4 7AL
U.K.

A modified Jacobi-Perron algorithm with explicitly given invariant measure

F. Schweiger

The theory of Jacobi-Perron algorithm can be viewed as the study of the transformation $T: B \to B$, defined as

$$T(x_1,x_2,\ldots,x_n) = \left(\frac{x_2}{x_1} - \left[\frac{x_2}{x_1} \right], \frac{x_3}{x_1} - \left[\frac{x_3}{x_1} \right], \ldots, \frac{1}{x_1} - \left[\frac{1}{x_1} \right] \right)$$

Here, B denotes the n-dimensional unit cube where a set of measure zero has been removed from such that all iterates of T are well defined. It is well known (see e.g. Schweiger [2]) that T is ergodic with respect to Lebesgue measure λ and that T admits a finite invariant measure μ, equivalent to λ. Since T is ergodic, μ is unique up to a multiplicative constant. If ρ denotes the density we may write

$$\mu(E) = \int_E \rho \, d\lambda$$

Only in the case $n = 1$ the "explicit shape" of ρ is known, namely

$$\rho(x) = \frac{1}{1+x} .$$

In this note we like to show that for a modified algorithm a similar formula can be given.

Let $D = \{ x \in B \mid 0 < x_i < x_1 < 1, \ 2 \le i \le n \}$.

Let S_n denote the symmetric group and Z_n the cyclic subgroup, generated by $(12\ldots n)$. Then for any $z = (z_1,\ldots,z_n) \in B$ there is a unique permutation $\sigma \in Z_n$ such that $\sigma z = (z_{\sigma 1},\ldots,z_{\sigma n}) \in D$. Note that the sets $\sigma^{-1}D$, $\sigma \in Z_n$, form a partition of B. Then we consider the transformation $S: D \to D$, defined by $S = \sigma T$, namely:

$$S(x_1,\ldots,x_n) = \sigma \left(\frac{x_2}{x_1}, \frac{x_3}{x_1}, \ldots, \frac{1}{x_1} - \left[\frac{1}{x_1} \right] \right).$$

This algorithm is similar to those considered by M.S.Waterman [3]. If $B(b) = \{ (x_1,x_2,\dots,x_n) \in D \mid [\frac{1}{x_1}] = b \}$ then $TB(b) = B$. Therefore $B(b)$ splits up into n sets $B(b,\sigma)$, $\sigma \in Z_n$, such that $SB(b,\sigma) = D$. Let $\kappa = (12\dots n) \in Z_n$, then the map

$$(x_1,x_2,\dots,x_n) \mapsto \left(\frac{1}{b+x_{\pi 1}}, \frac{x_{\pi 2}}{b+x_{\pi 1}}, \dots, \frac{x_{\pi n}}{b+x_{\pi 1}} \right)$$

maps D onto $B(b,\sigma)$ iff $\sigma\pi\kappa = \varepsilon$ (the identity in Z_n).

One can prove that S is ergodic and admits an invariant measure ν. Let ψ denote its density then ψ satisfies the Kuzmin's functional equation

$$\psi(x_1,x_2,\dots,x_n) = \sum_{b=1}^{\infty} \sum_{\sigma \in Z_n} \psi\left(\frac{1}{b+x_{\sigma 1}}, \frac{x_{\sigma 2}}{b+x_{\sigma 1}}, \dots, \frac{x_{\sigma n}}{b+x_{\sigma 1}}\right) \cdot \frac{1}{(b+x_{\sigma 1})^{n+1}}$$

<u>Theorem</u>: The function ψ is given up to a multiplicative constant by

$$\psi(x_1,x_2,\dots,x_n) = \sum_{\pi \in S_n} \frac{1}{1+x_{\pi 1}} \cdot \frac{1}{1+x_{\pi 1}+x_{\pi 2}} \cdots \frac{1}{1+x_{\pi 1}+\dots+x_{\pi n}}$$

<u>Proof</u>: In the sequel Σ_j means the sum over all permutations $\pi \in S_n$ such that $\pi j = 1$. Then

$$\psi\left(\frac{1}{b+x_1}, \frac{x_2}{b+x_1}, \dots, \frac{x_n}{b+x_1}\right) \frac{1}{(b+x_1)^{n+1}} =$$

$$= \Sigma_1 \frac{1}{b+x_1} \cdot \frac{1}{b+1+x_1} \cdot \frac{1}{b+1+x_1+x_{\pi 2}} \cdots \frac{1}{b+1+x_1+x_{\pi 2}+\dots+x_{\pi n}} +$$

$$+ \Sigma_2 \frac{1}{b+x_1} \cdot \frac{1}{b+x_1+x_{\pi 1}} \cdot \frac{1}{b+1+x_1+x_{\pi 1}} \cdots \frac{1}{b+1+x_1+x_{\pi 1}+\dots+x_{\pi n}} +$$

$+ \dots\dots +$

$$+ \Sigma_n \frac{1}{b+x_1} \cdot \frac{1}{b+x_1+x_{\pi 1}} \; \frac{1}{b+x_1+x_{\pi 1}+x_{\pi 2}} \cdots \frac{1}{b+1+x_1+x_{\pi 1}+\ldots+x_{\pi(n-1)}} =$$

$$= \Sigma_1 \Big(\frac{1}{b+x_1} \cdot \frac{1}{b+1+x_1+x_{\pi 2}} \cdots \frac{1}{b+1+x_1+x_{\pi 2}+\ldots+x_{\pi n}} -$$

$$- \frac{1}{b+1+x_1} \cdot \frac{1}{b+1+x_1+x_{\pi 2}} \cdots \frac{1}{b+1+x_1+x_{\pi 2}+\ldots+x_{\pi n}} \Big) +$$

$$+ \Sigma_2 \Big(\frac{1}{b+x_1} \cdot \frac{1}{b+x_1+x_{\pi 1}} \cdots \frac{1}{b+1+x_1+x_{\pi 1}+\ldots+x_{\pi n}} -$$

$$- \frac{1}{b+x_1} \cdot \frac{1}{b+1+x_1+x_{\pi 1}} \cdots \frac{1}{b+1+x_1+x_{\pi 1}+\ldots+x_{\pi n}} \Big) + \ldots\ldots +$$

$$+ \Sigma_n \Big(\frac{1}{b+x_1} \cdot \frac{1}{b+x_1+x_{\pi 1}} \cdots \frac{1}{b+x_1+x_{\pi 1}+\ldots+x_{\pi(n-1)}} -$$

$$- \frac{1}{b+x_1} \cdot \frac{1}{b+x_1+x_{\pi 1}} \cdots \frac{1}{b+1+x_1+x_{\pi 1}+\ldots+x_{\pi(n-1)}} \Big) =$$

$$= \Sigma_1 \Big(\frac{1}{b+x_{\tau 1}} \cdot \frac{1}{b+x_{\tau 1}+x_{\tau 2}} \cdot \frac{1}{b+x_{\tau 1}+x_{\tau 2}+x_{\tau 3}} \cdots \frac{1}{b+x_{\tau 1}+x_{\tau 2}+\ldots+x_{\tau n}} -$$

$$- \frac{1}{b+1+x_{\tau 1}} \cdot \frac{1}{b+1+x_{\tau 1}+x_{\tau 2}} \cdot \frac{1}{b+1+x_{\tau 1}+x_{\tau 2}+x_{\tau 3}} \cdots \frac{1}{b+1+x_{\tau 1}+x_{\tau 2}+\ldots+x_{\tau n}} \Big)$$

Therefore

$$\sum_{b=1}^{\infty} \psi\Big(\frac{1}{b+x_1}, \frac{x_2}{b+x_1}, \ldots, \frac{x_n}{b+x_1} \Big) \frac{1}{(b+x_1)^{n+1}} =$$

$$= \Sigma_1 \frac{1}{1+x_{\tau 1}} \cdot \frac{1}{1+x_{\tau 1}+x_{\tau 2}} \cdots \frac{1}{1+x_{\tau 1}+x_{\tau 2}+\ldots+x_{\tau n}}$$

Since S_n is just the complex product of Z_n by the set $\{ \tau \in S_n | \tau 1 = 1 \}$, we apply $\sigma \in Z_n$ on both sides. Then we sum over all $\sigma \in Z_n$ and obtain the result.

Remark: Podsypanin [1] considers the following related transformation $T^{\#}: B \to B$, defined for $n = 2$ as

$$T^{\neq}(x_1,x_2) = \left(\frac{x_2}{x_1}, \frac{1}{x_1} - \left[\frac{1}{x_1} \right] \right), \text{ if } x_2 \leq x_1$$

$$T^{\neq}(x_1,x_2) = \left(\frac{1}{x_2} - \left[\frac{1}{x_2} \right], \frac{x_1}{x_2} \right), \text{ if } x_1 < x_2.$$

It is easily checked that ψ is the density of an invariant measure with respect to $T^{\neq}$.

References

[1] PODSYPANIN, E.V.: A generalization of the continued fraction algorithm that is related to the Viggo Brun algorithm (Russian). Studies in Number Theory (LOMI), 4. Zap.Naučn. Sem. Leningrad. Otdel. Mat. Inst. Steklov 67 (1977), 184-194, 227.

[2] SCHWEIGER, F.: The metrical theory of Jacobi-Perron algorithm. Lecture Notes in Mathematics No. 334. Berlin-New York 1973.

[3] WATERMAN, M.S.: A Jacobi Algorithm and metric theory for greatest common divisors. J. Math. Anal. App. 59 (1977), 288-300.

F. Schweiger
Institut für Mathematik
Universität Salzburg
Petersbrunnstr. 19
A-5020 Salzburg

Ergodic Properties of Real Transformations

M. Thaler

In this note we shall present some remarks on the connection between the ergodicity and the rate of growth of the digits of certain number-theoretical transformations. We will consider transformations which describe algorithms with increasing sequences of digits. Classical examples are Engel series and Sylvester series. Some further examples will be quoted in the sequel.

A large number of these algorithms leads to transformations $T: (1,\infty) \to (1,\infty)$ of the following shape:

$T|_{(n,n+1)}: (n,n+1) \to (f(n), \infty)$ $(n \in \mathbb{N})$ is monotone and onto, where the map $f: (1,\infty) \to \mathbb{R}^+$ is increasing and continuous and fulfils $f(x) \geq x$. Let us recall the usual definition of the digits:

$$k_n(x) = m \iff T^{n-1}x \in B(m) = (m,m+1).$$

Owing to our assumptions we have

$$k_{n+1}(x) \geq f(k_n(x)) \geq k_n(x) \quad (\forall n).$$

So the function f determines the minimal rate of growth of the digits. In this paper ergodicity means ergodicity with respect to the Lebesgue measure. The domain B of all transformations in question is $(1,\infty)$ up to a set of measure zero. It is not difficult to see, that these transformations have no finite invariant measures absolutely continuous with respect to the Lebesgue measure. Roughly speaking, the conclusion of this note will be the following: If the function f grows rapidly enough, then T is not ergodic. However there is at least a special class of ergodic transformations of this form.

I) Non - ergodic cases

The most prominent among these are the Sylvester series:

$$Tx = \frac{(n+1)x}{n+1-x}, \quad x \in B(n),$$

and the cotangent algorithm:

$$Tx = \frac{nx+1}{x-n}, \quad x \in B(n).$$

It is well known that both are not ergodic. (For these we refer to Vervaat [7] and Schweiger [5]). The idea of the proof, used by Vervaat and Schweiger applies to a wider class of transformations. Sufficient for the applicability of this method are the following two conditions:

(i) There exists a constant C, such that

$$\frac{\lambda(B(k_1,\dots,k_n))}{\lambda(B(k_1,\dots,k_{n-1}))} \leq C \cdot \frac{f(k_{n-1})}{k_n^{\,2}}$$

for all admissible sequences (k_j) and for all $n \in \mathbb{N}$.

(ii) $f(n) \sim n^{1+\varepsilon}$ $(\varepsilon > 0)$

($B(k_1,\dots,k_t)$ denotes the cylinder of rank t based on the block $(k_1,\dots,k_t)$). The construction of an invariant set is as follows. Let g be the inverse of f, g^n the nth iterate of g and

$$y_n(x) = g^n(k_n(x)) \quad (x \in B).$$

From $k_{n+1}(x) \geq f(k_n(x))$ immediately follows

$$y_{n+1}(x) \geq y_n(x) \quad (x \in B).$$

An application of the lemma of Borel-Cantelli gives

$$\lim_{n\to\infty} y_n(x) = y(x) < \infty \quad \text{a.e.}$$

Now let I be a subset of $\mathbb{R}^+$ and $J(I) = \bigcup_{n \in \mathbb{Z}} f^n(I)$, then

$E(I) = \{x \in B\colon y(x) \in J(I)\}$ is invariant, since $y(Tx) = f(y(x))$. Some further information on the distribution of y allows to choose a set I such that $\lambda(E(I)) > 0$ and $\lambda(B \setminus E(I)) > 0$.

II) Ergodic cases

Examples of ergodic transformations of this type are the following:

A) $Tx = \frac{n+a}{x-n} + b, \quad x \in B(n)$

B) $Tx = \frac{n+a}{n+1-x} + b, x \in B(n)$,

where $a,b \in \mathbb{Z}$, $a \geq 0$, $a + b = c \geq 0$.

Let us mention some special cases known from the literature:

A) $a = b = 0$ .... F.Ryde [4]

$a = 0, b = 1$ .... T.A.Pierce [3]

$a = 1, b = 0$ .... F.Schweiger [6]

B) $a = 1, b = -1$ .... Engel series ([1], [2])

In case A) as well as in case B) we have

(a) $f(n) = n + c$

(b) $|V(n)'(x)| = \frac{n+a}{(x-b)^2}$, where $V(n)$

denotes the inverse of $T|_{(n,n+1)}$. The essential step in the proof of the ergodicity is the following

Lemma:

Let T satisfy (A) and (B) and let E be an invariant set. Then for $\alpha_n = \lambda(B(n) \cap E)$

$$| \alpha_n - \alpha_{n+1} | = O\left(\frac{1}{n^2}\right)$$

holds true.

Proof:

$$\Delta_n = \alpha_{n+1} - \alpha_n = \int_{n+c+1}^{\infty} \frac{n+a+1}{(x-b)^2} \cdot 1_E(x)dx - \int_{n+c}^{\infty} \frac{n+a}{(x-b)^2} \cdot 1_E(x)dx =$$

$$= \int_{n+c+1}^{\infty} \frac{1}{(x-b)^2} \cdot 1_E(x)dx - \int_{n+c}^{n+c+1} \frac{n+a}{(x-b)^2} \cdot 1_E(x)dx$$

As can be seen easily

$$\frac{\alpha_{n+c}}{(n+a+1)^2} \leq \int_{n+c}^{n+c+1} \frac{1}{(x-b)^2} 1_E(x)dx \leq \frac{\alpha_{n+c}}{(n+a)^2},$$

which yields

$$(1) \qquad \Delta_n \leq \int_{n+c+1}^{\infty} \frac{1}{(x-b)^2} 1_E(x)dx - \frac{n+a}{(n+a+1)^2} \int_{n+2c}^{\infty} \frac{n+c+a}{(x-b)^2} 1_E(x)dx$$

$$(2) \qquad \Delta_n \geq \int_{n+c+1}^{\infty} \frac{1}{(x-b)^2} 1_E(x)dx - \frac{1}{n+a} \int_{n+2c}^{\infty} \frac{n+c+a}{(x-b)^2} 1_E(x)dx$$

Let us first assume that $c \geq 1$ and let

$$\beta_n = \left| 1 - \frac{(n+a)(n+c+a)}{(n+a+1)^2} \right| .$$

Then, as a consequence of (1),

$$\Delta_n \leq \int_{n+c+1}^{n+2c} \frac{dx}{(x-b)^2} + \beta_n \int_{n+2c}^{\infty} \frac{dx}{(x-b)^2} \leq \frac{K_1}{n^2} .$$

Starting with (2) a similar calculation shows $\Delta_n \geq -\frac{K_2}{n^2}$. By an obvious modification we obtain the result for the case $c = 0$. □

The remaining arguments in the proof of the ergodicity are standard: Our lemma implies the existence of $\lim_n \alpha_n$.

Let us first consider the case when $\lim_n \alpha_n > 0$. Then there exists a $\delta > 0$ such that $\alpha_n \geq \delta$ for all n. A straight forward calculation shows

$$(\neq) \quad \sup_{x \in B(i)} \omega(k_1,\ldots,k_n)(x) \leq M \inf_{x \in B(i)} \omega(k_1,\ldots,k_n)(x)$$

for $i \geq k_n + c$, where M is a constant and $\omega(k_1,\ldots,k_n)$ is defined by

$$\int_F \omega(k_1,\ldots,k_n)(x)dx = \lambda(B(k_1,\ldots,k_n) \cap T^{-n}F).$$

Thus $\lambda(B(k_1,\ldots,k_n) \cap E) = \sum_{i \geq k_n+c} \int_{B(i) \cap E} \omega(k_1,\ldots,k_n) \geq$

$\geq \frac{\delta}{M} \cdot \lambda(B(k_1,\ldots,k_n))$ whenever $B(k_1,\ldots,k_n) \neq \emptyset$.

Hence $E = B$ (mod o). (We note that in case B) ($\neq$) does not hold if $a = b = 0$. However it is true under the assumption $k_1 \geq 2$. Therefore in this case the same arguments only gives $\lambda((2,\infty) \setminus E) = 0$. But this is sufficient. The exceptional behavior is caused by the fact, that $x = 1$ is a critical fixed point for T.) If $\lim_n \alpha_n = 0$ we apply the above arguments to $B \setminus E$ and obtain $\lambda(E) = 0$.

Remarks: The first proof for the ergodicity of Engel's transformation stems from L.Berg [1]. For this case and for the algorithm of T.A.Pierce [3] a very short proof has been given by Schweiger [5]. His method is based on the fact, that these transformations are conjugate to piecewise linear transformations on the unit interval. As our result demonstrates, this is not essential for the ergodicity.

The following two questions may be of interest:

1) Can one decide on the border line between ergodic and non-ergodic transformations of this type ?

2) What is the limiting distribution of the sequence of the random variables y_n considered in I) ? (In the case of Sylvester series Galambos [2] has proved, that the distribution function of y is continuously differentiable.)

References:

[1] Berg L.: Allgemeine Kriterien zur Maßbestimmung linearer Punktmengen. Math.Nachrichten 14 (1955), 263-285.

[2] Galambos J.: Representations of Real Numbers by Infinite Series. Lecture notes in Mathematics 502, Springer-Verlag, Berlin-Heidelberg-New York 1976.

[3] Pierce, T.A.: On an algorithm and its use in approximating roots of an algebraic equation. Amer.Math.Monthly 36(1929), 523-525.

[4] Ryde F.: Eine neue Art monotoner Kettenbruchentwicklungen. Ark.Mat. 1 (1951), 319-339.

[5] Schweiger F.: Lectures on fibered systems. Manuscript, Salzburg 1977.

[6] Schweiger F.: Eine kotangensalgorithmusähnliche Abbildung. J. Reine Angew.Math. 274/275 (1975), 90-93.

[7] Vervaat, W.: Success epochs in Bernoulli trials with applications in number theory. Math.Centre Tracts 42, Amsterdam 1972.

Dr.Maximilian Thaler

Mathematisches Institut
der Universität Salzburg
Petersbrunnstraße 19
A-5020 Salzburg/Österreich

Vol. 640: J. L. Dupont, Curvature and Characteristic Classes. X, 175 pages. 1978.

Vol. 641: Séminaire d'Algèbre Paul Dubreil, Proceedings Paris 1976-1977. Edité par M. P. Malliavin. IV, 367 pages. 1978.

Vol. 642: Theory and Applications of Graphs, Proceedings, Michigan 1976. Edited by Y. Alavi and D. R. Lick. XIV, 635 pages. 1978.

Vol. 643: M. Davis, Multiaxial Actions on Manifolds. VI, 141 pages. 1978.

Vol. 644: Vector Space Measures and Applications I, Proceedings 1977. Edited by R. M. Aron and S. Dineen. VIII, 451 pages. 1978.

Vol. 645: Vector Space Measures and Applications II, Proceedings 1977. Edited by R. M. Aron and S. Dineen. VIII, 218 pages. 1978.

Vol. 646: O. Tammi, Extremum Problems for Bounded Univalent Functions. VIII, 313 pages. 1978.

Vol. 647: L. J. Ratliff, Jr., Chain Conjectures in Ring Theory. VIII, 133 pages. 1978.

Vol. 648: Nonlinear Partial Differential Equations and Applications, Proceedings, Indiana 1976-1977. Edited by J. M. Chadam. VI, 206 pages. 1978.

Vol. 649: Séminaire de Probabilités XII, Proceedings, Strasbourg, 1976-1977. Edité par C. Dellacherie, P. A. Meyer et M. Weil. VIII, 805 pages. 1978.

Vol. 650: C*-Algebras and Applications to Physics. Proceedings 1977. Edited by H. Araki and R. V. Kadison. V, 192 pages. 1978.

Vol. 651: P. W. Michor, Functors and Categories of Banach Spaces. VI, 99 pages. 1978.

Vol. 652: Differential Topology, Foliations and Gelfand-Fuks-Cohomology, Proceedings 1976. Edited by P. A. Schweitzer. XIV, 252 pages. 1978.

Vol. 653: Locally Interacting Systems and Their Application in Biology. Proceedings, 1976. Edited by R. L. Dobrushin, V. I. Kryukov and A. L. Toom. XI, 202 pages. 1978.

Vol. 654: J. P. Buhler, Icosahedral Golois Representations. III, 143 pages. 1978.

Vol. 655: R. Baeza, Quadratic Forms Over Semilocal Rings. VI, 199 pages. 1978.

Vol. 656: Probability Theory on Vector Spaces. Proceedings, 1977. Edited by A. Weron. VIII, 274 pages. 1978.

Vol. 657: Geometric Applications of Homotopy Theory I, Proceedings 1977. Edited by M. G. Barratt and M. E. Mahowald. VIII, 459 pages. 1978.

Vol. 658: Geometric Applications of Homotopy Theory II, Proceedings 1977. Edited by M. G. Barratt and M. E. Mahowald. VIII, 487 pages. 1978.

Vol. 659: Bruckner, Differentiation of Real Functions. X, 247 pages. 1978.

Vol. 660: Equations aux Dérivée Partielles. Proceedings, 1977. Edité par Pham The Lai. VI, 216 pages. 1978.

Vol. 661: P. T. Johnstone, R. Paré, R. D. Rosebrugh, D. Schumacher, R. J. Wood, and G. C. Wraith, Indexed Categories and Their Applications. VII, 260 pages. 1978.

Vol. 662: Akin, The Metric Theory of Banach Manifolds. XIX, 306 pages. 1978.

Vol. 663: J. F. Berglund, H. D. Junghenn, P. Milnes, Compact Right Topological Semigroups and Generalizations of Almost Periodicity. X, 243 pages. 1978.

Vol. 664: Algebraic and Geometric Topology, Proceedings, 1977. Edited by K. C. Millett. XI, 240 pages. 1978.

Vol. 665: Journées d'Analyse Non Linéaire. Proceedings, 1977. Edité par P. Bénilan et J. Robert. VIII, 256 pages. 1978.

Vol. 666: B. Beauzamy, Espaces d'Interpolation Réels: Topologie et Géometrie. X, 104 pages. 1978.

Vol. 667: J. Gilewicz, Approximants de Padé. XIV, 511 pages. 1978.

Vol. 668: The Structure of Attractors in Dynamical Systems. Proceedings, 1977. Edited by J. C. Martin, N. G. Markley and W. Perrizo. VI, 264 pages. 1978.

Vol. 669: Higher Set Theory. Proceedings, 1977. Edited by G. H. Müller and D. S. Scott. XII, 476 pages. 1978.

Vol. 670: Fonctions de Plusieurs Variables Complexes III, Pro ceedings, 1977. Edité par F. Norguet. XII, 394 pages. 1978.

Vol. 671: R. T. Smythe and J. C. Wierman, First-Passage Perculatio on the Square Lattice. VIII, 196 pages. 1978.

Vol. 672: R. L. Taylor, Stochastic Convergence of Weighted Sum of Random Elements in Linear Spaces. VII, 216 pages. 1978.

Vol. 673: Algebraic Topology, Proceedings 1977. Edited by P. Ho man, R. Piccinini and D. Sjerve. VI, 278 pages. 1978.

Vol. 674: Z. Fiedorowicz and S. Priddy, Homology of Classic Groups Over Finite Fields and Their Associated Infinite Loop Space VI, 434 pages. 1978.

Vol. 675: J. Galambos and S. Kotz, Characterizations of Probabili Distributions. VIII, 169 pages. 1978.

Vol. 676: Differential Geometrical Methods in Mathematical Physics Proceedings, 1977. Edited by K. Bleuler, H. R. Petry and A. Reet VI, 626 pages. 1978.

Vol. 677: Séminaire Bourbaki, vol. 1976/77, Exposés 489-506. I 264 pages. 1978.

Vol. 678: D. Dacunha-Castelle, H. Heyer et B. Roynette. Ecole d'E de Probabilités de Saint-Flour. VII-1977. Edité par P. L. Hennequi IX, 379 pages. 1978.

Vol. 679: Numerical Treatment of Differential Equations in Applica tions, Proceedings, 1977. Edited by R. Ansorge and W. Törnig. I 163 pages. 1978.

Vol. 680: Mathematical Control Theory, Proceedings, 1977. Edite by W. A. Coppel. IX, 257 pages. 1978.

Vol. 681: Séminaire de Théorie du Potentiel Paris, No. 3, Directeurs M. Brelot, G. Choquet et J. Deny. Rédacteurs: F. Hirsch et G. Mokc bodzki. VII, 294 pages. 1978.

Vol. 682: G. D. James, The Representation Theory of the Symmetr Groups. V, 156 pages. 1978.

Vol. 683: Variétés Analytiques Compactes, Proceedings, 197 Edité par Y. Hervier et A. Hirschowitz. V, 248 pages. 1978.

Vol. 684: E. E. Rosinger, Distributions and Nonlinear Partial Di ferential Equations. XI, 146 pages. 1978.

Vol. 685: Knot Theory, Proceedings, 1977. Edited by J. C. Hausmann VII, 311 pages. 1978.

Vol. 686: Combinatorial Mathematics, Proceedings, 1977. Edite by D. A. Holton and J. Seberry. IX, 353 pages. 1978.

Vol. 687: Algebraic Geometry, Proceedings, 1977. Edited by L. D Olson. V, 244 pages. 1978.

Vol. 688: J. Dydak and J. Segal, Shape Theory. VI, 150 pages. 197

Vol. 689: Cabal Seminar 76-77, Proceedings, 1976-77. Edited b A.S. Kechris and Y. N. Moschovakis. V, 282 pages. 1978.

Vol. 690: W. J. J. Rey, Robust Statistical Methods. VI, 128 pages. 197

Vol. 691: G. Viennot, Algèbres de Lie Libres et Monoïdes Libres III, 124 pages. 1978.

Vol. 692: T. Husain and S. M. Khaleelulla, Barrelledness in Topo logical and Ordered Vector Spaces. IX, 258 pages. 1978.

Vol. 693: Hilbert Space Operators, Proceedings, 1977. Edited b J. M. Bachar Jr. and D. W. Hadwin. VIII, 184 pages. 1978.

Vol. 694: Séminaire Pierre Lelong – Henri Skoda (Analyse) Anné 1976/77. VII, 334 pages. 1978.

Vol. 695: Measure Theory Applications to Stochastic Analysis Proceedings, 1977. Edited by G. Kallianpur and D. Kölzow. X 261 pages. 1978.

Vol. 696: P. J. Feinsilver, Special Functions, Probability Semigroups and Hamiltonian Flows. VI, 112 pages. 1978.

Vol. 697: Topics in Algebra, Proceedings, 1978. Edited by M. F. New man. XI, 229 pages. 1978.

Vol. 698: E. Grosswald, Bessel Polynomials. XIV, 182 pages. 197

Vol. 699: R. E. Greene and H.-H. Wu, Function Theory on Manifold Which Possess a Pole. III, 215 pages. 1979.